Nicolaos D. Epiotis

Theory of
Organic Reactions

With 69 Figures and 47 Tables

Springer-Verlag
Berlin Heidelberg New York 1978

Nicolaos Demetrios Epiotis

Assistant Professor of Chemistry
University of Washington
Seattle, WA 98195/USA

ISBN-3-540-08551-3 Springer-Verlag Berlin Heidelberg New York
ISBN-0-387-08551-3 Springer-Verlag New York Heidelberg Berlin

Library of Congress Cataloging in Publication Data. Epiotis, N D 1944– Theory of organic reactions. (Reactivity and structure ; v. 5) Bibliography: p. Includes indexes. 1. Chemical reactions. 2. Chemistry, Organic. I. Title. II. Series. QD501.E68   547′.1′39   77-17405

© by Springer-Verlag Berlin Heidelberg 1978
Printed in Germany

Typesetting: Elsner & Behrens, Oftersheim
Printing and binding: Konrad Triltsch, Würzburg
2152/3140-543210

To my parents

Prologue

"Apparently there is color, apparently sweetness, apparently bitterness, actually there are only atoms and the void." These words have been attributed to the Greek philosopher Democritus about 420 B.C.; he justifiably becomes the father of chemical theory. In 1812 Berzelius proposed that all chemical combinations are caused by electrostatic attractions. As we shall see, our work, 165 years thereafter, suggests exactly that, albeit in a sense different from the one implied by the originator of the idea. At the beginning of the 20th century Kössel and Lewis made important contributions toward the understanding of ionic and covalent bonds, respectively. These developments along with the ideas of van't Hoff and Le Bel regarding the shapes of organic molecules provided the impetus for a systematic investigation of structure and reactivity in organic chemistry. In 1927, Heitler and London first made use of quantum mechanics to give a valence bond description of the covalent bond. Subsequently, "resonance theory" [1], based on valence bond theory [2], became the organic chemists' favorite tool for rationalizing and predicting chemical behavior. This approach is still the mainstay of undergraduate education; its popularity is due not only to its potency as a theoretical tool but also to its conceptual accessibility to students and practitioners of organic chemistry who have had no exposure to formal quantum mechanics. By the turn of the second half of this century, Molecular Orbital (MO) theory began to attract the attention of organic chemists but never became an overpowering influence until the publication of the Woodward-Hoffmann rules in 1965 [3, 4]. The reasons for the failure of MO theory to replace "resonance theory" prior to that time are understandable. Specifically, many problems which were on the focus of interest up until 1965, such as the rate and orientation of aromatic substitution, the effect of substituents on the rate of a solvolytic reaction, etc., found a qualitative answer on the basis of "resonance theory". As a result, the success of MO theory in rationalizing the same set of facts was impressive but not compelling.

The publication of the Woodward-Hoffmann approach to the stereochemistry of pericyclic reactions [4 a] constitutes a milestone in the development of organic chemical theory for two reasons:

a) The Woodward-Hoffmann rules have correlated a great number of existing chemical facts and stimulated further widespread experimentation.

b) The Woodward-Hoffmann rules have demonstrated that intuition based on "resonance theory" is insufficient to deal with a set of unexpected phenomena such as molecular transformations involving non least motion pathways. In this sense, these rules have provided compelling testimony for the utility of MO theory in organic chemistry.

The success of the Woodward-Hoffmann formulation of the stereochemistry of pericyclic reactions appeared impressive and as a result led to the crystallization of certain viewpoints regarding chemical reactivity. By 1970, the vast majority of organic chemists subscribed to the dogma that Woodward-Hoffmann "allowed" reactions are facile, one step, reactions which can proceed with a high degree of stereoselectivity. On the other hand, Woodward-Hoffmann "forbidden" reactions were expected to proceed via well defined intermediates lacking pericyclic bonding and, thus, result in stereorandom product formation. Furthermore, the "forbidden" nature of a given reaction was thought to imply a large activation energy necessary for the conversion of reactants to products.

In 1968, I noticed that certain experimental results were in striking contrast to expectations based upon the Woodward-Hoffmann analysis. For example, thermal polar $2\pi + 2\pi$ cycloadditions were known to be extremely facile reactions proceeding in a highly stereoselective or stereospecific manner [5]. These and other experimental observations led to the suspicion that the electronic nature of reactants plays a pivotal role in determining the preferred reaction path, e.g., whatever is valid for the reaction of ethylene with ethylene may not necessarily be valid for the reaction of tetramethoxyethylene with tetracyanoethylene. Accordingly, I attempted to develop a general theory of chemical reactivity where the dependence of reaction rates upon the electronic properties of the reactants is treated explicitly. In one formalism, the wavefunction of the reaction complex was constructed from the reactant MO's [6]; in a second formalism, it was constructed from the MO's of the reaction "supermolecule" [7]. The objective has been to connect the shapes and energies of reactant MO's with the relative energies of stereochemically and regiochemically distinct reaction complexes. Since the effect of substitution upon the shapes and energies of the MO's of the parent system could be easily understood in terms of perturbation theory [8], the influence of substituents upon the relative stabilities of stereoisomeric and regioisomeric reaction complexes could be deduced on the basis of relatively simple arguments. The results of these analyses suggested an overall picture of chemical reactivity which was substantially different from the one currently accepted. Nonetheless, a key problem remained to be resolved before one could be convinced that this new viewpoint was not an artifact of the admittedly drastic simplifications adopted. This problem had to do with the validity of the static model approximation employed in our theoretical approach.

With the exception of the correlation diagram approach of Longuet-Higgins [4 b], Abrahamson [4 b], and Woodward and Hoffmann [4 a], most qualitative approaches to problems of chemical reactivity have made use of the static model approximation [4 d, 4 e, 4 f, 6, 7, 9 – 15] according to which two reactants are taken to interact at a fixed intermolecular distance and the resulting stabilization energy is assumed to be an index of the height of the barrier of the corresponding reaction. The success of this model in interpreting many puzzling reactivity trends provided grounds for optimism. On the other hand, its deficiencies are many and will be discussed in the main text. Accordingly, a qualitative theory of chemical reactivity which makes possible the construction of Potential Energy (P. E.) surfaces began to loom as a definite necessity.

In this work, a general theoretical procedure for constructing qualitative P. E. surfaces will be outlined and will be used to interpret organic reactivity. Thus, it would not be inaccurate to say that this monograph constitutes the beginning of a new conceptual approach to organic chemistry rather than the consolidation of known facts and familiar ideas. The key theoretical notions presented here no doubt will be further refined and elaborated; the P. E. surfaces will become more detailed and accurate as more information about the energy states of molecules become available; and probing experiments will be designed to resolve questions which cannot be answered at the level of theory presented here. The aim of this work is to stimulate interest in thermal and photochemical organic reactivity and to suggest ways in which gas phase and solution mechanistic chemistry, synthesis, spectroscopy and theory can be united. The objective is the mastery of the science and art of drawing the all important P. E. surfaces which reveal how reactants become products.

Contents

Contents

Contents

1. One-Determinental Theory of Chemical Reactivity

1.1 General Principles and Computational Schemes

Qualitative organic theory involves the reduction of quantum mechanical equations to simple concepts. We examine the sequential steps needed to make a transition from the ab initio Self Consistent Field MO (SCF MO) method to the qualitative One Electron MO (OEMO) model which currently constitutes the theoretical framework for most discussions of organic chemical problems.

The total molecular Hamiltonian can be displayed in the manner shown below, where τ are electronic and T nuclear spatial coordinates, $\hat{K}$ is the kinetic energy operator, and $\hat{V}$ is the potential energy operator.

$$\hat{H}_T = \hat{K}(\tau) + \hat{K}(T) + \hat{V}(\tau, T) + \hat{V}(T) \tag{1.1}$$

The Born-Oppenheimer adiabatic approximation allows us to write the molecular wavefunctions as follows:

$$X = \Theta\Psi \tag{1.2}$$

Θ is a nuclear eigenfunction of $\hat{K}(T) + E_{el}(T) + \hat{V}(T) + \langle\Psi \mid \hat{K}(T) \mid \Psi\rangle$, Ψ is an electronic eigenfunction of $\hat{K}(\tau) + \hat{V}(\tau, T)$, and $E_{el}(T)$ is the electronic energy for a given nuclear configuration. The former operator will be denoted by $\hat{H}_N$ and the latter by $\hat{H}$. The total energy can now be approximated as the sum of the electronic energy and the internuclear repulsive energy. The latter term can be calculated classically and the problem is reduced to the computation of the former term.

The electronic wavefunction Ψ should now be written in a manner which allows for the development of a simple conceptual picture of the electronic structure of the system in question. A simple way of meeting this requirement amounts to writing Ψ as a product of single electron functions, ω_ν, which describe the motion of electrons i.

$$\Psi(1, 2, \ldots n) = \omega_1(1) \, \omega_2(2) \ldots \omega_\nu(n) \tag{1.3}$$

This is an independent particle type wavefunction. Recognition of the fact that an electron has spin leads to replacement of the functions ω_ν by product functions.

$$\omega_\nu = \phi_\nu \sigma_\nu \tag{1.4}$$

Here, ϕ_ν is a function of the spatial coordinates, σ_ν is a function of the spin coordinates and ω_ν is termed a spin orbital (AO or MO). For the case of an electron, σ_ν can be α or β. Accordingly, we shall employ the following symbolism in all subsequent discussions:

$$\omega_\nu = \phi_\nu \alpha = \phi_\nu \tag{1.5}$$

$$\omega_\nu' = \phi_\nu \beta = \bar{\phi}_\nu \tag{1.6}$$

Furthermore, recognition of the fact that electrons are indistinguishable leads us to reconsider the form of the wavefunction as written in equation 1.3. An operation which interchanges any two electrons should have no effect on the physical properties of the system since electron labeling is a procedural task within the confines of a mathematical treatment. A wavefunction should be written in such a way that interchanging two electrons may lead to a change of sign but not magnitude. The Pauli principle states that an electronic wavefunction is antisymmetric for an interchange of any pair of electrons. For an n electron system, this is written as a Slater determinant of spin orbitals which satisfies the Pauli principle.

$$\Psi = \frac{1}{\sqrt{2n!}} \begin{vmatrix} \phi_1(1) & \phi_1(2) & \phi_1(3) \ldots \phi_1(2n) \\ \bar{\phi}_1(1) & \bar{\phi}_1(2) & \bar{\phi}_1(3) \ldots \bar{\phi}_1(2n) \\ \vdots & \vdots & \vdots \qquad \vdots \\ \bar{\phi}_n(1) & \bar{\phi}_n(2) & \bar{\phi}_n(3) \ldots \bar{\phi}_n(2n) \end{vmatrix} \tag{1.7}$$

A Slater determinant can be represented in a variety of ways, some of which are shown below:

$$\Psi = |\phi_1(1) \bar{\phi}_1(2) \ldots \bar{\phi}_n(2n)| = |\Psi_d| \tag{1.8}$$

$$\Psi = \hat{A}\Psi_d \tag{1.9}$$

$$\Psi = \frac{1}{\sqrt{2n!}} \sum_{\hat{P}} (-1)^p \hat{P}\Psi_d \tag{1.10}$$

In the above equations, Ψ_d is the diagonal term of the determinant, $\hat{A}$ is a normalized antisymmetrization operator, $\hat{P}$ is a permutation operator and p is the number of electron pair interchanges required to regenerate Ψ_d from $\hat{P}\Psi_d$.

Once we have specified the form of the electronic wavefunction Ψ and the form of the operator $\hat{H}$, we can write the following equation, where integration should be carried out over all spatial, $d\tau$, and spin, $d\sigma$, coordinates and Ψ is assumed to be real:

$$E = \frac{\int\int \Psi \hat{H} \Psi d\tau d\sigma}{\int\int \Psi^2 d\tau d\sigma} \tag{1.11}$$

For convenience, we can write $\hat{H}$ as a sum of one electron terms and two electron terms:

$$\hat{H} = \sum_i \hat{h}_i + \sum_{i<j}\sum \frac{e^2}{r_{ij}} \tag{1.12}$$

where

$$\hat{h}_i = -\frac{h^2}{8\pi^2 m} \nabla_i^2 - \sum_A^N \frac{e^2 Z_A}{r_{Ai}} \tag{1.13}$$

In shorthand, we can write

$$\hat{H} = \hat{O}_1 + \hat{O}_2 \tag{1.14}$$

where $\hat{O}_1$ is the one electron and $\hat{O}_2$ the two electron part of the operator.

For a closed shell molecule, the total electronic energy is given by the expression below:

$$E = 2\sum_\mu I_\mu + \sum_\mu \sum_\nu (2J_{\mu\nu} - K_{\mu\nu}) \tag{1.15}$$

where

$$I_\mu = \int \phi_\mu(i)\hat{h}(i)\phi_\mu(i) d\tau_i \tag{1.16}$$

$$J_{\mu\nu} = \int\int \phi_\mu(i)\phi_\mu(i) \frac{e^2}{r_{ij}} \phi_\nu(j)\phi_\nu(j) d\tau_i d\tau_j \tag{1.17}$$

$$K_{\mu\nu} = \int\int \phi_\mu(i)\phi_\nu(i) \frac{e^2}{r_{ij}} \phi_\mu(j)\phi_\nu(j) d\tau_i d\tau_j \tag{1.18}$$

The best orbitals ϕ_μ are those which minimize the energy and simultaneously satisfy the condition of orthogonality. By using well known procedures [16–18], we can arrive at the final equation shown below.

$$\left[\hat{h} + \sum_\nu (2\hat{J}_\nu - \hat{K}_\nu)\right] \phi_\mu = \sum_\nu \lambda_{\mu\nu}\phi_\nu \tag{1.19}$$

where

$$\hat{J}_\nu(i)\phi_\mu(j) = e^2 \left(\int \frac{\phi_\nu(i)\phi_\nu(i)}{r_{ij}} d\tau_i\right) \phi_\mu(j) \tag{1.20}$$

3

$$\hat{K}_\nu(i)\phi_\mu(j) = e^2 \left(\int \frac{\phi_\mu(i)\phi_\nu(i)}{r_{ij}} \, d\tau_i \right) \phi_\nu(j) \qquad (1.21)$$

The reader should notice that equation 1.19 does not have the usual form of an eigenvalue problem. The $\lambda_{\mu\nu}$ are constants (Langrangian multipliers) which can be chosen arbitrarily, subject to the orbital orthonormality condition. One such choice is the one which makes the λ matrix diagonal:

$$\lambda_{\mu\nu} = \epsilon_\mu \delta_{\mu\nu} \qquad (1.22)$$

This choice reduces equation 1.19 to the form:

$$\hat{H}^0 \phi_\mu = \epsilon_\mu \phi_\mu \qquad (1.23)$$

In the case of molecules, the MO's which satisfy equation 1.23 are called canonical MO's and they are eigenfunctions of a Hamiltonian which commutes with the molecular symmetry operators.

In this book, we shall be interested in reactions of molecules. Hence, we should consider the case where ϕ_μ is an MO. The most common representation of an MO involves expansion in terms of a complete set of one electron functions. These one electron functions x_k are usually AO's in order to lend a conceptual simplicity to the interpretation of the final result:

$$\phi_\mu = \sum_k c_k^\mu x_k \qquad (1.24)$$

The expansion coefficients serve as vehicles for introducing variations in the MO's.

The procedure for obtaining the solutions of equation 1.23 in terms of an AO basis set is simple. Firstly, we substitute 1.24 into 1.16, 1.17 and 1.18 in order to transform the I, J and K MO integrals into integrals over AO's.

$$I_\mu = \sum_k \sum_l c_k^\mu c_l^\mu h_{kl} \qquad (1.25)$$

$$J_{\mu\nu} = \sum_k \sum_l \sum_m \sum_n c_k^\mu c_l^\mu c_m^\nu c_n^\nu \langle kl \mid mn \rangle \qquad (1.26)$$

$$K_{\mu\nu} = \sum_k \sum_l \sum_m \sum_n c_k^\mu c_m^\mu c_l^\nu c_n^\nu \langle km \mid ln \rangle \qquad (1.27)$$

In the above expressions we have two kinds of integrals over AO's, namely, one electron integrals of the type h_{kl} and two electron integrals of the type $\langle kl \mid mn \rangle$ where

$$\langle kl \mid mn \rangle = \int \int x_k(1)x_l(1) \frac{e^2}{r_{12}} x_m(2)x_n(2)d\tau_1 d\tau_2 \qquad (1.28)$$

$$\langle km \mid ln \rangle = \int \int x_k(1)x_m(1) \frac{e^2}{r_{12}} x_l(2)x_n(2)d\tau_1 d\tau_2 \qquad (1.29)$$

4

The expressions 1.25, 1.26 and 1.27 are now substituted in equation 1.15 and we obtain E in terms of integrals over AO's.

$$E = \sum_{\mu} \left\{ \sum_{k} \sum_{l} c_k^\mu c_l^\mu h_{kl} + \sum_{\mu} \sum_{\nu > \mu} \sum_{k} \sum_{l} \sum_{m} \sum_{n} 2 c_k^\mu c_m^\nu \left[c_l^\mu c_n^\nu \langle kl \mid mn \rangle - 1/2 c_n^\mu c_l^\nu \langle km \mid ln \rangle \right] \right\}$$

$$(1.30)$$

Now, the best orbitals are those which minimize the energy subject to the condition that the normalization of the μth MO is retained. By using well known procedures [16–18], we obtain the secular equations

$$\sum_{l} (F_{kl} - S_{kl} \epsilon_\mu) c_l^\mu = 0 \quad \mu = 1, 2, 3 \ldots \qquad (1.31)$$

where

$$F_{kl} = h_{kl} + \sum_{\nu} \left\{ \sum_{m} \sum_{n} c_m^\nu c_n^\nu [2 \langle kl \mid mn \rangle - \langle km \mid ln \rangle] \right\} \qquad (1.32)$$

$$S_{kl} = \int x_k x_l d\tau \qquad (1.33)$$

$$\epsilon_\mu = \lambda_{\mu\mu} \qquad (1.34)$$

We are now prepared to have an overview of MO calculations in organic chemistry on the basis of the secular equations 1.31. The levels of computational sophistication are the following:

a) Ab initio SCF MO methods [17]. These methods involve the explicit computation of the h_{kl}, S_{kl} and $\langle kl \mid mn \rangle$ integrals and solution of the secular equations to yield the eigenvalues ϵ_μ and eigenvectors c_k^μ by an iterative approach. The latter is necessary because one needs the orbitals he is seeking to determine in order to construct the F matrix and solve the secular equations. Thus, an initial choice of orbitals is made, the F matrix is constructed, and the secular equations are solved to yield a first approximation of the orbitals. A new cycle is initiated and the procedure is repeated until convergence is achieved.

b) Neglect of Differential Overlap (NDO) SCF MO methods [18]. These methods employ the approximation $x_k x_l d\tau = 0$ applied to all or certain AO pairs. This approximation reduces the number of repulsion integrals which must be computed. The ones which survive the NDO approximation can be calculated explicitly or approximated empirically. Once more, an iterative solution of the secular equations leads to the desired eigenvalues and eigenvectors. Typical examples of such methods are the CNDO and CNDO/2 (Complete Neglect of Differential Overlap), INDO (Intermediate Neglect of Differential Overlap), etc., methods.

c) Hückel type methods which include overlap [19]. In such methods, the true Hamiltonian is replaced by an effective one electron Hamiltonian and the secular equations take the form

$$\sum_{l} (H'_{kl} - \epsilon_\mu S_{kl}) c_l^\mu = 0 \qquad\qquad (1.35)$$

The one electron matrix elements H'_{kl} are evaluated empirically, the overlap integrals are calculated explicitly or approximated empirically, and the secular equations can be solved noniteratively. A typical method of this type is the Extended Hückel (EH) method.

d) Hückel type methods which neglect overlap [20]. These methods differ from the previously mentioned EH methods insofar as the following approximation is made:

$$S_{kl} = \delta_{kl} \qquad\qquad (1.36)$$

Our discussion of the various computational methods in organic chemistry may lead someone to believe that, given the requisite computer facilities and a generous grant, ab initio calculations provide the best hope for understanding chemistry. However, the success of these methods depends upon the choice of the basis set, the nature of the experimental quantity which one seeks to determine, etc. By contrast, we have concentrated on the identification of factors which are responsible for chemical trends [21]. Once such a philosophy is adopted, calculations at the ab initio, NDO, or, EH level are performed not so much in order to obtain a number and compare it with an appropriate experimental result but, rather, in order to test a theoretical analysis of the type to be described below.

1.2 Qualitative One-Determinental Models of Chemical Reactivity

In attempting to develop a qualitative theory of chemical reactions, one has to choose a theoretical formalism as well as an appropriate model which allows for simplification of an inherently complex problem. In general, there are two types of models:

a) The static model. In this case, the energy change, ΔE, which accompanies the interaction of two molecules or molecular fragments is evaluated at a certain assumed intermolecular distance. This distance is kept fixed for the purpose of making comparisons between two or more reaction systems.

b) The dynamic model. In this case, the energy change which accompanies the interaction of two molecules or molecular fragments is evaluated along the entire reaction coordinate.

The interpretation of any calculations carried out within the static model approximation is possible only if certain assumptions are made. Specifically, the calculated $\Delta\Delta E$ at the assumed intermolecular distance may have the following meanings:

a) $\Delta\Delta E$ reflects relative energies of transition state complexes. In this case, the chosen intermolecular distance is appropriate to a transition state complex and, for comparative purposes, all such complexes are assumed to occur at the same intermolecular distance.

b) $\Delta\Delta E$ reflects relative magnitudes of initial slopes of P.E. surfaces and, thus, provides indirect information about relative energies of transition state complexes assuming that these occur at comparable intermolecular distances.

The static model breaks down if two transition states under comparison occur at significantly different intermolecular distances.

Once a model has been chosen, the next step involves the choice of the appropriate theoretical method which can be used to calculate the energy change accompanying the interaction of two molecules or molecular fragments. These methods have been described in the previous section.

1.3 The Static One Electron Molecular Orbital Model

The simplest theoretical approach to the qualitative interpretation of structural preferences and reactivity trends of organic molecules employs Hückel theory in the static model approximation. A Hückel calculation yields the energy and wavefunction of a total system designated A–B–C. However, greater insight into the chemical properties of the system is gained by the stepwise construction of A–B–C starting from the constituent parts, A, B, and C.

A   A–B

+ $\xrightarrow{\text{Step 1}}$ + $\xrightarrow{\text{Step 2}}$ A–B–C

B   C

The constituent parts A, B and C may be atoms or molecules and A–B–C a triatomic molecule or a termolecular transition state complex. The construction of the wavefunction of the total system can be achieved as long as the wavefunctions of the parts are known. In such treatments, we may distinguish two important types of orbital interactions:

a) Interaction of degenerate orbitals (or orbitals which are very close in energy).

b) Interaction of nondegenerate orbitals which are separated by a substantial energy gap.

In such problems, one makes use of an effective one electron Hamiltonian operator, $\hat{H}'$, and overlap may or may not be neglected. The interrelationships between the unperturbed orbitals of the fragments and the orbitals of the composite system resulting from the interactions of the unperturbed orbitals of the fragments are depicted in Scheme 1.

The above considerations lead directly to the formulation of a general theory for structure and reactivity. In particular, structural problems can be approached by considering a molecule as a composite system of fragments and reactivity problems can be approached by considering a reaction complex as a composite system of atoms or molecules. Then, chemical trends can be deduced from the following recipe:

a) Dissect a given molecule or reaction complex into fragments.

Scheme 1.

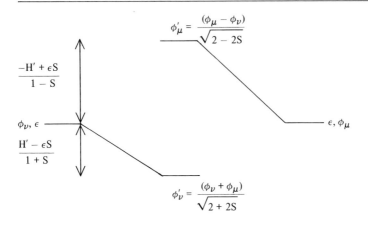

$$\phi'_\mu = \frac{(\phi_\mu - \phi_\nu)}{\sqrt{2 - 2S}}$$

$$\frac{-H' + \epsilon S}{1 - S}$$

$$\phi_\nu, \epsilon \qquad\qquad\qquad\qquad \epsilon, \phi_\mu$$

$$\frac{H' - \epsilon S}{1 + S}$$

$$\phi'_\nu = \frac{(\phi_\nu + \phi_\mu)}{\sqrt{2 + 2S}}$$

$$\phi'_\mu = \frac{\phi_\mu - \lambda_\mu \phi_\nu}{\sqrt{1 + 2\lambda_\mu S + \lambda_\mu^2}}$$

$$\frac{(H' - \epsilon_\mu S)^2}{\epsilon_\mu - \epsilon_\nu}$$

$$\epsilon_\mu, \phi_\mu$$

$$\phi_\nu, \epsilon_\nu \qquad\qquad\qquad\qquad \frac{(H' - \epsilon_\nu S)^2}{\epsilon_\nu - \epsilon_\mu}$$

$$\phi'_\nu = \frac{\phi_\nu + \lambda_\nu \phi_\mu}{\sqrt{1 + 2\lambda_\nu S + \lambda_\nu^2}}$$

Definitions: $\quad S = S_{\mu\nu} = \langle \phi_\mu | \phi_\nu \rangle$

$\qquad\qquad\quad H' = H'_{\mu\nu} = \langle \phi_\mu | \hat{H}' | \phi_\nu \rangle$

$\qquad\qquad\quad \lambda_x = \dfrac{H' - S\epsilon_x}{\epsilon_\nu - \epsilon_\mu} \quad (x = \mu, \nu)$

b) Identify the most important stabilizing or destabilizing orbital interaction(s) between the fragments. This procedural task is best conveyed by the construction of interaction diagrams which show the crucial MO's of the two fragments, their occupancy, and their interactions.

c) Investigate the conditions under which the dominant stabilizing or destabilizing orbital interaction(s) is (are) maximized.

d) Consider the possibility of additional important stabilizing or destabilizing inter-actions competing with the principal stabilizing or destabilizing interaction(s). If this materializes, some "quantitative" calculation of the relative magnitudes of the various interactions becomes desirable.

8

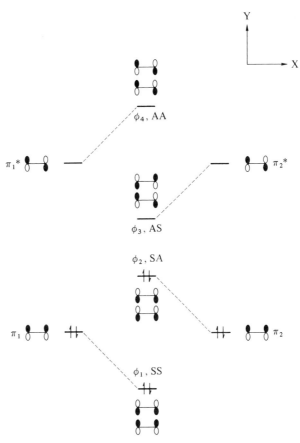

Fig. 1. MO interaction diagram for $_2\pi_s + _2\pi_s$ union of two ethylenes. Each MO of the reaction complex is classified by reference to the ZY and ZX symmetry planes, in order. Overlap is neglected

This qualitative model, which makes use of an effective one electron Hamiltonian, will be designated henceforth the One Electron MO (OEMO) model.

We shall now discuss an application of the OEMO model to a typical chemical problem. For example, consider the $2\pi + 2\pi$ cycloaddition of two ethylenes which can proceed via two different pathways, one designated $_2\pi_s + _2\pi_s$ and the other $_2\pi_s + _2\pi_a$. The relative energies of the corresponding reaction complexes can be ascertained from consideration of the dominant pi MO interactions which obtain in the two cases. The interaction diagrams of Figs. 1 and 2 clearly show that there is greater stabilization and smaller destabilization in the case of $_2\pi_s + _2\pi_a$. Accordingly, this pathway is predicted to be favored *electronically*[1] over the $_2\pi_s + _2\pi_s$ pathway, i.e., the relative stabilization

---

1   In this treatise, electronic effects are related to the interaction of AO's, MO's, or configurations with respect to a monoelectronic Hamiltonian. On the other hand, steric effects, are related to interelectronic and internuclear repulsive interactions.

energies of the two reaction complexes are assumed to be indices of the relative barrier heights involved in the two pathways.

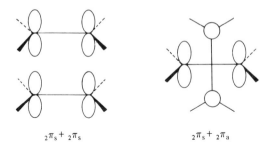

$_2\pi_s + _2\pi_s$ $_2\pi_s + _2\pi_a$

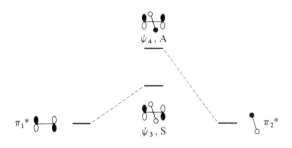

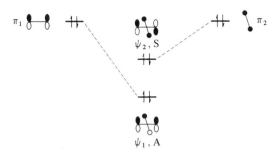

Fig. 2. MO interaction diagram for $_2\pi_s + _2\pi_a$ union of two ethylenes. Each MO of the reaction complex is classified by reference to a twofold symmetry axis. Overlap is neglected

1.4 Orbital Energies. The Donor-Acceptor Classification of Molecules and the Concept of Reaction Polarity

The identification of the dominant MO interaction presupposes knowledge of the magnitudes of energy gaps and the sizes of interaction matrix elements. Accordingly, it is necessary to delineate a procedure for the evaluation of these quantities.

Orbital energies, and, hence, energy gaps, can be obtained from explicit quantum mechanical calculations or experimental ionization potentials, electron affinities and electronic transitions. The relationship between experimental data and orbital energies is defined by the following equations where I stands for the ionization potential, A for the electron affinity, and G for the energy of the electronic transition from the Highest Occupied (HO) MO to the Lowest Unoccupied (LU) MO.

$$\epsilon(HO) \simeq -I(HO) \tag{1.37}$$

$$\epsilon(LU) \simeq A(LU) \tag{1.38}$$

or

$$\epsilon(LU) \simeq -I(HO) + G(HO \rightarrow LU) \tag{1.39}$$

In addition, orbital energies can be used to define the acid-base interrelationship of any two molecules or fragments. Thus, in considering a pair of reactants, we may designate one partner as the *donor* and the other as the *acceptor* in such a manner so that the following condition is satisfied:

$$I_D - A_A < I_A - A_D \tag{1.40}$$

In the above inequality, I_D is the ionization potential of the donor, I_A the ionization potential of the acceptor, A_D the electron affinity of the donor, and A_A the electron affinity of the acceptor. In terms of the corresponding quantities of MO theory and when both partners have a closed shell electronic configuration, the condition becomes:

$$\left| \epsilon_{HO}^D - \epsilon_{LU}^A \right| < \left| \epsilon_{HO}^A - \epsilon_{LU}^D \right| \tag{1.41}$$

Typical examples of donor-acceptor pairs are given below, where R is a pi electron releasing group and W a pi electron withdrawing group.

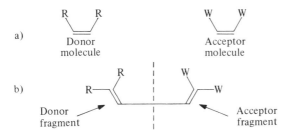

11

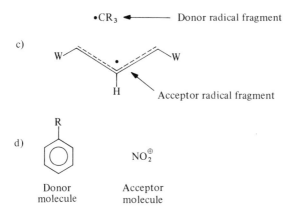

c)

•CR$_3$ ◄——— Donor radical fragment

Acceptor radical fragment

d)

Donor
molecule

Acceptor
molecule

In general, a high energy HO is the earmark of a good donor and a low energy LU that of a good acceptor. R type substituents such as OH, SH, NH_2, PH_2, etc. will tend to make a molecule a good donor, W type substituents such as CHO, CN, etc., will tend to make a molecule a good acceptor, and U type substituents such as $CH=CH_2$, C_6H_5, etc., will tend to make a molecule simultaneously a good donor and a good acceptor.

The preceding discussion sets the stage for the definition of the all important term *reaction polarity,* or, simply, *polarity,* P, which will be encountered very often in the subsequent chapters. This term refers to the donor-acceptor properties of a pair of reactants and is quantitatively defined by the following equation:

$$P = (I_D - A_A)^{-1} \tag{1.42}$$

1.5 One Electron Interaction Matrix Elements and Overlap Integrals

We now turn our attention to interaction matrix elements, $H'_{\mu\nu}$. The simplest approximation of $H'_{\mu\nu}$ is the one shown below.

$$H'_{\mu\nu} = kS_{\mu\nu} \tag{1.43}$$

In equation 1.43, k is an energy constant, and $S_{\mu\nu}$ an MO overlap integral. This approximation is valid if the overlapping centers involve the same atoms in two cases being compared. If this condition is not met, $H_{\mu\nu}$ is expanded in terms of interaction matrix elements over AO's, γ_{mn}, and each of these can be approximated as follows:

$$\gamma_{mn} = k'(\beta_A^0 + \beta_B^0) S_{mn} \tag{1.44}$$

In equation 1.44, k' is a numerical constant, β_A^0 is a quantity related to the average orbital ionization energy of A, and S_{mn} an AO overlap integral.

An extensive discussion of one electron interaction matrix elements can be found in Epiotis *et al.* [21]. We assume here that either approximation of $H'_{\mu\nu}$ leads to qualitatively similar results, unless otherwise specified.

The ability to evaluate MO overlap integrals and, as a result, one electron interaction matrix elements, is an important prerequisite for the understanding of our approach. Hence, we provide a sample illustration by considering the MO overlap integral between the pi HO of ethylene and the pi LU of butadiene in the geometry shown below.

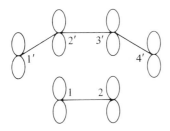

First, HO and LU are replaced by their Linear Combination of Atomic Orbitals (LCAO) expressions.

$$HO = a_1 p_1 + a_2 p_2 \tag{1.45}$$

$$LU = b_1 p'_1 - b_2 p'_2 - b_3 p'_3 + b_4 p'_4 \tag{1.46}$$

Accordingly, we have:

$$S_{HO,LU} = [a_1 b_1 \langle p_1 | p'_1 \rangle + a_2 b_4 \langle p_2 | p'_4 \rangle] + [a_1 b_4 \langle p_1 | p'_4 \rangle + a_2 b_1 \langle p_2 | p'_1 \rangle$$
$$- a_1 b_2 \langle p_1 | p'_2 \rangle - a_2 b_2 \langle p_2 | p'_2 \rangle - a_1 b_3 \langle p_1 | p'_3 \rangle - a_2 b_3 \langle p_2 | p'_3 \rangle] \tag{1.47}$$

The magnitude of the overlap integral will depend on two factors:

a) The magnitude of each of the coefficient products $a_m b_n$.

b) The magnitude of each AO overlap integral S_{mn} which will increase as the intermolecular distance decreases.

One can differentiate between small and large contributions. The large contribution is made up by the terms in the first bracket of equation 1.47 which involve AO overlap integrals between the primarily overlapping AO's i.e., $p_1 - p'_1$ and $p_2 - p'_4$. The smaller contribution is made up by the terms in the second bracket which involve AO overlap integrals between secondarily overlapping AO's, i.e., $p_1 - p'_2$, $p_1 - p'_3$, $p_1 - p'_4$, $p_2 - p'_3$, $p_2 - p'_2$, and $p_2 - p'_1$. In many instances, the primary contribution to the MO overlap integral remains constant for two geometries but the secondary contribution varies. In such cases, secondary orbital overlap effects may exert a profound influence on the preferred reaction path.

A classification of MO overlap integrals, which will be important in future discussions of chemical reactivity, is based upon the *symmetry* of the overlapping MO's. Using Greek capital letters in order to symbolize AO coefficients and referring to the drawing below, we distinguish the following three cases:

a) The overlapping MO's have the same symmetry or pseudosymmetry. In this case, $S_{\mu\nu} \propto (A\Gamma + A\Gamma)s$ (symmetric case) or $S_{\mu\nu} \propto (A\Gamma + B\Delta)s$ (pseudosymmetric case).

b) The overlapping MO's have different symmetry or pseudosymmetry. In this case, $S_{\mu\nu} \propto (A\Gamma - A\Gamma)s$ (symmetric case) or $S_{\mu\nu} \propto (A\Gamma - B\Delta)s$ (pseudosymmetric case).

c) The overlapping MO's have different pseudosymmetry and B or Δ equals zero. In this case, $S_{\mu\nu} \propto (A\Gamma)s$.

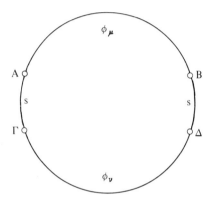

In chemical reactivity, the size of $H'_{\mu\nu}$ and, hence, $S_{\mu\nu}$, as well as the energy gap $\epsilon_\mu - \epsilon_\nu$, control the magnitude of a key orbital interaction. In most cases, as the electronic nature of one reactant is changed by substitution, the quantities $H'_{\mu\nu}$ and $\epsilon_\mu - \epsilon_\nu$ change in such a manner so that the variation of $H'_{\mu\nu}$ may favor stronger interaction while the variation of $\epsilon_\mu - \epsilon_\nu$ may operate in an opposite direction. For example, consider the reaction of 1,3-butadiene and each of the members of the series $H_2C=CH_2$, $(CN)CH=CH_2$, $(CN)_2C=CH_2$, $(CN)_2C=CH(CN)$, and $(CN)_2C=C(CN)_2$ to yield a cyclohexene derivative. Focusing on the interaction between the HO of the donor diene and the LU of the acceptor dienophile, $|\epsilon_\mu - \epsilon_\nu|$ decreases because the LU of the dienophile progressively decreases in energy and so does $H'_{\mu\nu}$ in terms of its absolute magnitude. This arises because the electron density of the dienophile LU at the centers of primary overlap becomes smaller due to the sequential replacement of hydrogens by cyano groups.

The "dilution" of the LU electron density of the dienophile and, in more general terms, the "dilution" of orbital electron density as a result of increasing substitution leads to progressively less negative values of $H'_{\mu\nu}$. This effect has very important consequences for chemical reactivity which will be discussed later. As an example, the LU electron density at vinylic carbon of ethylene is 0.50, while that of tetracyanoethylene (TCNE) is 0.25 (CNDO/2 calculation) [18a].

1.6 The Woodward-Hoffmann Molecular Orbital Correlation Diagram

Currently, the most popular application of Hückel theory in the dynamic model approximation is the MO correlation diagram approach proposed by Woodward and Hoffmann in their treatment of pericyclic reactions. According to this formulism, the energy change of a given electronic configuration of the reaction system along the reaction coordinate is evaluated by reference to the energy changes of the MO's. At the level of Hückel theory, the energy change of an MO which accompanies a perturbation of one Coulomb integral (α_m) and one resonance integral (β_{mn}) is given by the equation

$$\delta E^\mu = q^\mu_m \delta \alpha_m + 2p^\mu_{mn} \delta \beta_{mn} \tag{1.48}$$

where q^μ_m is the electron density of the mth AO of the μth MO, p^μ_{mn} the bond order between the mth and nth AO's of the μth MO, $\delta \alpha_m$ the Coulomb integral perturba-

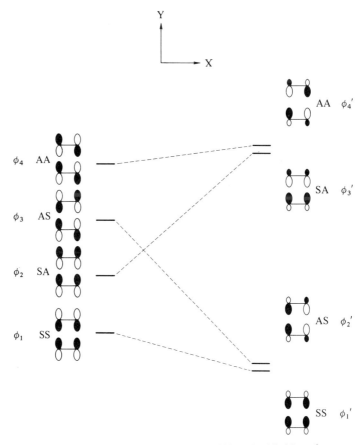

Fig. 3. MO correlation diagram for $_2\pi_s + {}_2\pi_s$ cycloaddition. Each MO is classified by reference to the ZY and ZX symmetry planes, in order

tion of the mth AO, and $\delta\beta_{mn}$ the resonance integral perturbation between the mth and nth AO's. The total energy change of the system is given, in Hückel theory, by

$$\delta E_T = \sum_\mu \delta E^\mu = \sum_\mu q_m^\mu \, \delta\alpha_m + 2 \sum_\mu p_{mn} \delta\beta_{mn} \tag{1.49}$$

If more than one α_m and β_{mn} change, the previous expressions become

$$\delta E^\mu = \sum_m q_m^\mu \delta\alpha_m + 2 \sum_{m<n} \sum_m p_{mn}^\mu \delta\beta_{mn} \tag{1.50}$$

and

$$\delta E_T = \sum_\mu \left(\sum_m q_m^\mu \delta\alpha_m \right) + \sum_\mu \left(2 \sum_{m<n} \sum_m p_{mn}^\mu \delta\beta_{mn} \right) \tag{1.51}$$

Now, for a given reaction system we can assume that $\delta\alpha_m = 0$ and that the resonance integral between two AO's is proportional to their overlap, S_{mn}. Hence, equation 1.51 becomes

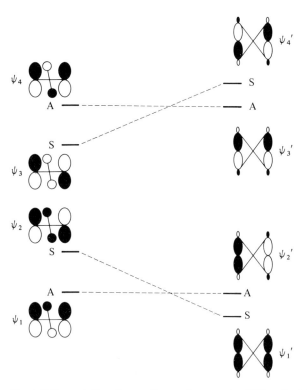

Fig. 4. MO correlation diagram for $_2\pi_s + {}_2\pi_a$ cycloaddition. Each MO is classified by reference to a twofold symmetry axis

$$\delta E_T \propto - \sum_{\mu} \left(\sum_{m<n} \sum_{m} p_{mn}^{\mu} \delta S_{mn} \right) \qquad (1.52)$$

The latter equations tells us that the energy of a given configuration increases as the overlap between two AO's interacting in a bonding manner (i.e. $p_{mn}^{\mu} > 0$) decreases, or, the overlap between two AO's interacting in an antibonding manner (i.e. $p_{mn}^{\mu} < 0$) increases, and conversely.

A good illustration of the MO correlation diagram approach is provided by the ethylene cycloaddition discussed before in connection with the static OEMO method. The pertinent diagrams are given in Figs. 3 and 4. The total energy of the system, which in Hückel theory is a simple sum of occupied orbital energies, rises (overlap included) or remains constant (overlap neglected) during the initial phase of $_2\pi_s + _2\pi_s$ addition and before the MO crossing point is reached. On the other hand, it constantly decreases in the case of $_2\pi_s + _2\pi_a$ addition. Hence, one can infer that the latter pathway will be more favorable.

2. Configuration Interaction Overview of Chemical Reactivity

2.1 General Principles

The current qualitative theories of organic chemistry are based on one-determinental MO theory. In such approaches one constructs the MO's of a total system from the MO's of fragments and assumes that the energy of the configuration generated by this procedure approximates the energy of the corresponding state. This is a reasonable expectation for most thermal reactions but a rather unrealistic one for photochemical reactions. Adopting a better approach, one seeks to generate electronic states from electronic configurations, which, in turn, have been generated from fragment MO's. This is the CI approach. In this chapter, two different CI methods which can be utilized in the analysis of actual chemical problems will be described. As background, a brief outline of the theoretical manipulations involved in a typical CI calculation follows.

In the simplest case, the steps which one has to follow in going from MO's to configurations and, finally, to states are the following:

a) The Born-Oppenheimer approximation is made.

b) The molecular Hamiltonian operator can be written as in equation 1.12.

c) The MO's of the system are calculated using one of the theoretical methods described in Chap. 1.

d) The MO's are used to construct the spin and symmetry adapted basis configurations. These may involve a single determinant, or, a sum of determinants.

e) The electronic states of the system are calculated. In the simplest case of a basis set of two configurations Ψ_ρ and Ψ_σ, one must solve the equations shown below in matrix form.

$$\begin{bmatrix} \langle \Psi_\rho | \hat{H} | \Psi_\rho \rangle - E & \langle \Psi_\sigma | \hat{H} | \Psi_\rho \rangle \\ \langle \Psi_\rho | \hat{H} | \Psi_\sigma \rangle & \langle \Psi_\sigma | \hat{H} | \Psi_\sigma \rangle - E \end{bmatrix} \begin{bmatrix} C_1 \\ C_2 \end{bmatrix} = 0 \qquad (2.1)$$

We use the following notation:

$$\langle \Psi_\rho | \hat{H} | \Psi_\sigma \rangle = H_{\rho\sigma} \tag{2.2}$$

Furthermore, we assume:

$$H_{\rho\sigma} = H_{\sigma\rho} \tag{2.3}$$

$$\begin{bmatrix} H_{\rho\rho} & H_{\rho\sigma} \\ H_{\rho\sigma} & H_{\sigma\sigma} \end{bmatrix} \begin{bmatrix} C_1 \\ C_2 \end{bmatrix} = \begin{bmatrix} E & 0 \\ 0 & E \end{bmatrix} \begin{bmatrix} C_1 \\ C_2 \end{bmatrix} \tag{2.4}$$

The 2 x 2 matrix on the left hand side of equation 2.4 is termed the energy matrix and we are now faced with the task of evaluating matrix elements between determinental wavefunctions. In general, these matrix elements can be written as follows:

$$H_{\rho\sigma} = \langle \Psi_\rho | \hat{H} | \Psi_\sigma \rangle \tag{2.5}$$

$$H_{\rho\sigma} = \langle \hat{A}\Psi_{d,\rho} | \hat{H} | \hat{A}\Psi_{d,\sigma} \rangle \tag{2.6}$$

$$H_{\rho\sigma} = \sum_{\hat{P}} (-1)^p \langle \Psi_{d,\rho} | \hat{H} | \hat{P}\Psi_{d,\sigma} \rangle \tag{2.7}$$

where Ψ_ρ and Ψ_σ are normalized Slater determinants. Now, the various matrix elements can be expressed in terms of MO integrals with respect to the one electron part ($\hat{O}_1$) and the two electron part ($\hat{O}_2$) of the Hamiltonian.

$$\langle \phi_\mu(i) | \hat{O}_1 | \phi_\nu(i) \rangle = I_{\mu\nu} \tag{2.8}$$

$$\langle \phi_\mu(i)\phi_\nu(j) | \hat{O}_2 | \phi_\kappa(i)\phi_\lambda(j) \rangle = \langle \mu\kappa | \nu\lambda \rangle \tag{2.9}$$

The well known Coulomb ($J_{\mu\nu}$) and exchange ($K_{\mu\nu}$) integrals are special cases of the general repulsion integral ($\mu\kappa | \nu\lambda$).

2.2 Qualitative Configuration Interaction Models of Chemical Reactivity

As in the case of one-determinental theory, one can use the following two models in order to develop a qualitative CI approach:

a) The static model. The electronic states of reaction complexes are calculated using a fixed intermolecular distance. The state energy is usually expressed in the form of a stabilization (or destabilization) energy of the reaction complex relative to the non-interacting components.

b) The dynamic model. The electronic states of the reaction complex are calculated for intermolecular distances varying from infinity to the final bonding distance. This leads to the construction of P.E. surfaces for ground and excited state reactions. Transition states and intermediates can readily be identified and the relative energetics of different reaction pathways can be assessed.

2.3 The Static Linear Combination of Fragment Configurations Method

Molecules and reaction complexes can be viewed as composite systems made up of two (or more) atomic or molecular fragments. Unless the two fragments are identical, we can define one as the donor fragment **D** and the other as the acceptor fragment **A**. The electronic states of the composite system **D**–**A** at a chosen intermolecular distance can be described in terms of a *Linear Combination of Fragment Configurations* (LCFC) method [22]. A complete knowledge of the eigenvalues and eigenvectors of the states of **D**–**A** for any geometrical arrangement of the nuclei of the constituent atoms leads to a good understanding of the chemical behavior of the system.

The various steps involved in a static LCFC analysis are the following:

a) The zero order configurations are generated by permuting all electrons of **D** and **A** among the MO's of **D** and **A**.

b) The zero order configurations are spin and symmetry adapted to yield the basis configurations.

c) The energy matrix is constructed and the elements are computed at a fixed interfragment distance *neglecting interfragment differential MO overlap, i.e.* $\phi^D \psi^A d\tau = 0$. *This approximation will be referred to as the ZIDMOO* (Zero Interfragment Differential MO Overlap) *approximation.*

d) The configurations are ranked according to their relative energies.

e) The final electronic states are generated by assuming that the interaction of two nondegenerate configurations is proportional to the square of the interaction matrix element and inversely proportional to their energy separation.

As an illustration, consider the interaction of a no bond configuration (DA) which involves open shell components with higher energy charge transfer configurations (D^+A^- and D^-A^+) which involve closed shell components. The zero order configurations and the associated energy terms are shown in Fig. 5. The ZIDMOO approximation "turns off" the exchange stabilization, $2\beta S$ but preserves the CI stabilization of DA. Both effects depend on β. This is a typical situation: Specifically, open shell configurations which are parents of the ground and first excited states enjoy exchange stabilization which depends on the same β as CI stabilization due to interaction with an upper, close lying configuration. Furthermore, exchange destabilization, $-2\beta S$, may contribute to the energies of closed *and* open shell configurations, in general. This effect, which disappears in the ZIDMOO approximation, contributes to steric repulsion which can be treated in an empirical manner. These observations constitute the simplest justification of the ZIDMOO approximation in comparative studies of chemical reactivity (see also Chap. 22).

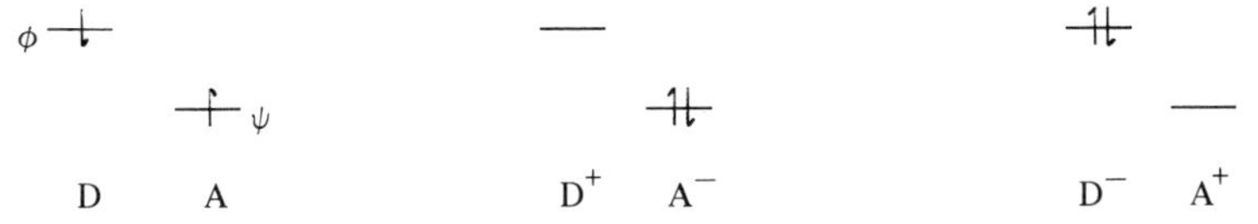

$\phi \dashv\vdash$ $\dashv$ $\dashv\vdash$

   $\dashv\vdash \psi$ $\dashv\vdash$ $\dashv$

  D   A D^+ A^- D^- A^+

Energy Term	LCFC	LCFC-ZIDMOO
$\langle DA\|\hat{H}\|DA\rangle$	$\dfrac{1}{1+S^2}(\epsilon_\psi + \epsilon_\phi + J_{\psi\phi} + V_a + V_b + 2\beta S + K_{\psi\phi}) + V_{nn}$	$\epsilon_\psi + \epsilon_\phi + J_{\psi\phi} + V_a + V_b + V_{nn}$
$\langle D^+A^-\|\hat{H}\|D^+A^-\rangle$	$2\epsilon_\psi + 2V_a + J_{\psi\psi} + V_{nn}$	$2\epsilon_\psi + 2V_a + J_{\psi\psi} + V_{nn}$
$\langle D^-A^+\|\hat{H}\|D^-A^+\rangle$	$2\epsilon_\phi + 2V_b + J_{\phi\phi} + V_{nn}$	$2\epsilon_\phi + 2V_b + J_{\phi\phi} + V_{nn}$
$\langle DA\|\hat{H}\|D^+A^-\rangle$	$\sqrt{\dfrac{2}{1+S^2}}\left[\beta + \langle\psi\phi\|\dfrac{1}{r_{12}}\|\psi\psi\rangle + (\epsilon_\psi + V_a)S\right]$	$\sqrt{2}\beta$
$\langle DA\|\hat{H}\|D^-A^+\rangle$	$\sqrt{\dfrac{2}{1+S^2}}\left[\beta + \langle\psi\phi\|\dfrac{1}{r_{12}}\|\phi\phi\rangle + (\epsilon_\phi + V_b)S\right]$	$\sqrt{2}\beta$
$\langle D^+A^-\|\hat{H}\|D^-A^+\rangle$	$2\beta S + K_{\psi\phi}$	0

Definitions

		(1)	(2)
S	$= \langle i\|j\rangle$		
ϵ_i	$= \langle i\| -\dfrac{1}{2}\nabla^2 - \dfrac{Z_a}{r_{1a}}\|i\rangle$	$\bullet$ a	$\bullet$ b
J_{ij}	$= \langle ii\|\dfrac{1}{r_{12}}\|jj\rangle$	a, b = Nuclei	
K_{ij}	$= \langle ij\|\dfrac{1}{r_{12}}\|ij\rangle$	1, 2 = Electrons	
V_b	$= \langle i\| -\dfrac{Z_b}{r_{1b}}\|i\rangle$		
V_{nn}	$= \dfrac{Z_a Z_b}{r_{ab}}$		
β	$= \langle i\| -\dfrac{1}{2}\nabla^2 - \dfrac{Z_a}{r_{1a}} - \dfrac{Z_b}{r_{1b}}\|j\rangle$		

Fig. 5. Rigorous and approximate energies of typical DA, D^+A^- and D^-A^+ configurations

A further simplification of the static LCFC analysis involves the following steps:

a) Construction of the zero order as well as the spin and symmetry adapted configurations as indicated before.

b) Ranking the various configurations according to their relative energies by reference to the following equations:

$$\epsilon(DA) = 0 \tag{2.10}$$

$$\epsilon(D^+A^-) = I_D - A_A + C \tag{2.11}$$

$$\epsilon(D^-A^+) = I_A - A_D + C' \tag{2.12}$$

where I_D is the ionization potential of the donor, A_A is the electron affinity of the acceptor, C is the coulombic attractive term, etc. The relative energies of higher excited configurations can be obtained by addition of appropriate excitation energies to (2.10)–(2.12).

c) The interaction matrix elements over configurations, $\langle \Psi_\rho | \hat{P} | \Psi_\sigma \rangle$, are set proportional to appropriately corresponding MO overlap integrals, $S_{\mu\mu'}$[23]. This amounts to saying that the interaction Hamiltonian is a Hückel-type one electron Hamiltonian. The MO overlap integral is the quantity which provides the basis for the connection of MO symmetry and sizes of AO coefficients with chemical reactivity.

The evaluation of one electron interaction matrix elements between two determinental wavefunctions is very simple [24]. For example, consider the interaction matrix element of the DA and D^+A^- zero order configurations shown below in the ZIDMOO approximation. The two MO's which differ in occupancy by one electron in the two configurations are ϕ and χ. Hence, the interaction matrix element is proportional to $S_{\phi,\chi}$.

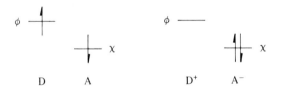

From now on, we shall use the symbol $\mu-\mu'$ to designate the MO overlap integral $S_{\mu\mu'}$, dropping the preceeding sign and constant for brevity wherever appropriate.

The form of the interaction matrix element, i.e., the form of the corresponding MO overlap integral, is extremely important because it can be connected with important reaction aspects. Two types of reactions can be distinguished:

a) Bicentric reactions involving primary interaction between one center of one reactant with one center of the second reactant. In this case, $\hat{P}$ operates along two centers and $S_{\mu\mu'}$ can be zero or nonzero. When $S_{\mu\mu'} = 0$, Ψ_ρ and Ψ_σ do not mix and constitute *nonpericyclic* states (Fig. 6a), when $S_{\mu\mu'} \neq 0$, Ψ_ρ and Ψ_σ mix and give rise to one *pericyclic* and one *antipericyclic* state (Fig. 6b).

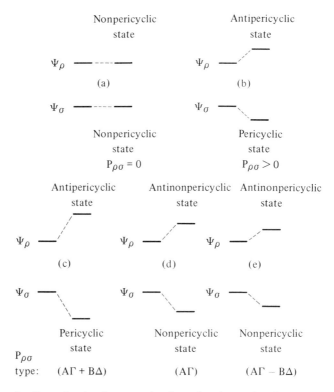

Fig. 6. The generation of pericyclic, antipericyclic, nonpericyclic, and antinonpericyclic states. (a–b) Bicentric reactions. (c–e) Multicentric reactions

b) Multicentric reactions involving primary interaction of two centers of one reactant with one or two centers of the second reactant. In this instance, $\hat{P}$ operates along two pairs of interacting centers, e.g., pair A and pair B. Thus, we may write:

$$\hat{P} = \hat{P}_A + \hat{P}_B \tag{2.13}$$

The matrix element of two configurations Ψ_ρ and Ψ_σ with respect to $\hat{P}$ can be written as follows:

$$\langle \Psi_\rho | \hat{P} | \Psi_\sigma \rangle = \langle \Psi_\rho | \hat{P}_A | \Psi_\sigma \rangle + \langle \Psi_\rho | \hat{P}_B | \Psi_\sigma \rangle \tag{2.14}$$

We distinguish the following cases of interest:

1. $\langle \Psi_\rho | \hat{P}_A | \Psi_\sigma \rangle$ and $\langle \Psi_\rho | \hat{P}_B | \Psi_\sigma \rangle$ have the same sign. This situation arises when $S_{\mu\mu'}$ is of the general type $(A\Gamma + B\Delta)$s. $\langle \Psi_\rho | \hat{P} | \Psi_\sigma \rangle$ is large and the two configurations mix strongly due to bonding interactions along both pairs of atoms. The states arising from such mixing will be *pericyclic* and *antipericyclic* states (see Fig. 6c).

2. $\langle \Psi_\rho | \hat{P}_A | \Psi_\sigma \rangle$ and $\langle \Psi_\rho | \hat{P}_B | \Psi_\sigma \rangle$ have opposite signs. This situation arises when $S_{\mu\mu'}$ is of the general type $(A\Gamma - B\Delta)$s. $\langle \Psi_\rho | \hat{P} | \Psi_\sigma \rangle$ is very small and the two configura-

23

tions mix slightly due to a bonding and an antibonding interaction along the two pairs of atoms. The states arising from such mixing will be *nonpericyclic* and *antinonpericyclic* states (see Fig. 6 e).

3. $\langle \Psi_\rho | \hat{P}_A | \Psi_\sigma \rangle$ is nonzero and $\langle \Psi_\rho | \hat{P}_B | \Psi_\sigma \rangle$ is zero. This situation arises when $S_{\mu\mu'}$ is of the general type $(A\Gamma)$s. $\langle \Psi_\rho | \hat{P} | \Psi_\sigma \rangle$ is still appreciable and the two configurations mix due to a bonding and a nonbonding interaction along the two pairs of atoms. The states arising from such mixing will be *nonpericyclic* and *antinonpericyclic* states (see Fig. 6 d).

c) The final electronic states are generated by adhering to the rule that when two non-degenerate configurations interact their energy change will be inversely proportional to the energy difference between the two configurations and directly proportional to the square of the interaction matrix element. When two degenerate configurations interact, their energy change will be directly proportional to the interaction matrix element.

As an illustration, the static LCFC approach can be applied to the problem of $2\pi + 2\pi$ cycloaddition of two ethylenes. The zero order configurations are shown in Fig. 7. Proper symmetry adaptation requires that we take appropriate linear combinations of the degenerate configuration wave functions. Thus, we have:

$$\Psi_1 = D_1 D_2 \tag{2.15}$$

$$\Psi_2^+ = \frac{1}{\sqrt{2}} \ [(D_1^+ D_2^-) + (D_1^- D_2^+)] \tag{2.16}$$

$$\Psi_2^- = \frac{1}{\sqrt{2}} \ [(D_1^+ D_2^-) - (D_1^- D_2^+)] \tag{2.17}$$

$$\Psi_3^+ = \frac{1}{\sqrt{2}} \ [(D_1^* D_2) + (D_1 D_2^*)] \tag{2.18}$$

$$\Psi_3^- = \frac{1}{\sqrt{2}} \ [(D_1^* D_2) - (D_1 D_2^*)] \tag{2.19}$$

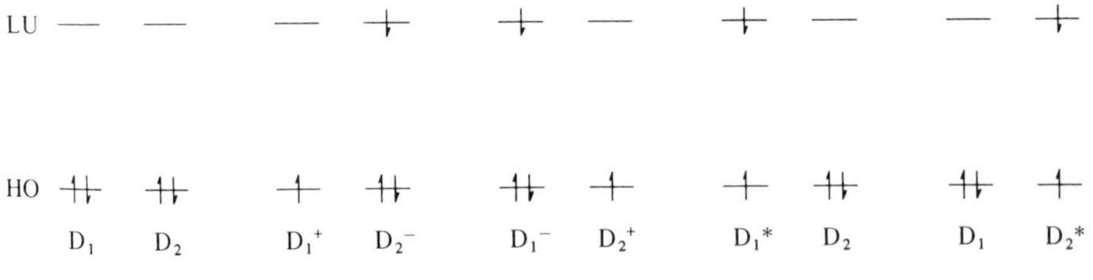

Fig. 7. Zero order configurations for the static LCFC treatment of the $2\pi + 2\pi$ ethylene cyclodimerization

24

Recalling that two zero order configurations which differ by two spin orbitals cannot be mixed via the Hamiltonian employed in our analysis, the energies of the basic configurations are as follows:

$$\epsilon(\Psi_1) = 0 \tag{2.20}$$

$$\epsilon(\Psi_2^+) = \epsilon(\Psi_2^-) = \epsilon(D_1^+ D_2^-) = I_{D_1} - A_{D_2} + C \tag{2.21}$$

$$\epsilon(\Psi_3^+) = \epsilon(\Psi_3^-) = \epsilon(D_1^* D_2) = G(HO \rightarrow LU) \tag{2.22}$$

The interaction matrix is given in Table 1.

Table 1. Interaction matrix for $2\pi + 2\pi$ cycloaddition of two ethylenes

	Ψ_1	Ψ_2^+	Ψ_2^-	Ψ_3^+	Ψ_3^-
Ψ_1	–	$(HO^{D_1}-LU^{D_2})$ $+(HO^{D_2}-LU^{D_1})$	0	0	0
Ψ_2^+		–	0	0	0
Ψ_2^-			–	0	$(HO^{D_1}-HO^{D_2})$ $+(LU^{D_1}-LU^{D_2})$
Ψ_3^+				–	0
Ψ_3^-					–

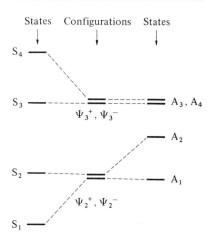

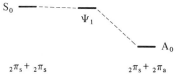

Fig. 8. Static LCFC diagram for $2\pi_s + 2\pi_s$ and $2\pi_s + 2\pi_a$ ethylene cyclodimerization. The electronic states for the former case are symbolized by S_n and those for the latter case by A_n

25

The static LCFC diagram is shown in Fig. 8. In the case of reactions occurring via the $_2\pi_s + _2\pi_s$ complex, the DA cannot interact with any of the higher lying configurations, and thus, there is no stabilization of the corresponding pathway. By contrast, in the case of a reaction proceeding via the $_2\pi_s + _2\pi_a$ complex, DA interacts with Ψ_2^+ and the corresponding pathway is stabilized. Accordingly, the thermal reaction is expected to proceed via the latter path because A_0 has lower energy than S_0. By following the same line of reasoning, we can predict that the photochemical reaction will proceed via the $_2\pi_s + _2\pi_s$ path because S_1 has now lower energy than A_1.

2.4 The Static Delocalized Configuration Interaction Method

The Delocalized CI (DCI) approach is fundamentally different from the one previously described in terms of the nature of the zero order basis set configurations. We designate the two fragments which make up the molecule or transition state as **D** and **A**. The electronic states of the composite system **D**–**A** at a chosen intermolecular distance can now be described in terms of a linear combination of monoexcited, diexcited, etc., basis configurations which, in contrast to the LCFC approach, involve MO's delocalized over both **D** and **A** fragments. The fully delocalized MO's can be approximately derived by OEMO theory from the MO's of the fragments **D** and **A**. By means of this approach, the electronic nature of the fragments is expressed in the final MO manifold of the composite system **D**–**A**. In turn, the properties of this MO manifold determine the extent of interaction between the various basis configurations to produce the final electronic states.

The procedure outlined above does not allow one to make a simple operational distinction between different electronic types of complexes involved in thermal and photochemical reactions because the MO's extend over both fragments and this precludes an immediate identification of states which have a high contribution of local excitation (expressed in the LCFC approach by the D*A and DA* configurations) or charge transfer (expressed in the LCFC approach by the D^+A^- and D^-A^+ configurations).

Thermal reactions can be analyzed qualitatively by means of the static DCI approach using a minimal basis set comprised of the ground and lowest diexcited configurations (Ψ^0 and $\Psi_{\kappa\rightarrow\tau}^2$, respectively). The diagonal elements of the energy matrix are the energies of the two wavefunctions. Since we are interested only in relative energies, we can write:

$$\epsilon^0 = 0 \tag{2.23}$$

$$\epsilon_{\kappa\rightarrow\tau}^2 = 2(I_\tau - I_\kappa) - \left\{ \sum_{\nu=1}^2 [(2J_{\kappa\nu} - K_{\kappa\nu}) - (2J_{\nu\tau} - K_{\nu\tau})] + J_{\tau\tau} - 2J_{\kappa\tau} + K_{\kappa\tau} \right\} \tag{2.24}$$

26

where I is the energy of an MO with respect to the one electron part of the Hamiltonian and J and K are the usual Coulomb and exchange integrals. For many chemical situations, the variation of the term in braces is small compared to the variation of $2(I_\tau - I_\kappa)$ and, since we are interested in qualitative trends, we can write

$$\epsilon^0 = 0 \tag{2.25}$$

$$\epsilon^2_{\kappa \to \tau} \simeq 2(I_\tau - I_\kappa) \tag{2.26}$$

The off-diagonal elements of the energy matrix are simply:

$$\langle \Psi^0 | \hat{H} | \Psi^2_{\kappa \to \tau} \rangle = \langle \Psi^2_{\kappa \to \tau} | \hat{H} | \Psi^0 \rangle = K_{\kappa\tau} \tag{2.27}$$

The eigenvalues of the eigenstates resulting from the mixing of Ψ^0 and $\Psi^2_{\kappa \to \tau}$ can be found variationally in the usual manner. If $\epsilon^2_{\kappa \to \tau} - \epsilon^0$ is large compared to the matrix element $K_{\kappa\tau}$, we can obtain the solutions in the usual perturbation form (neglecting overlap):

$$E^0 = \epsilon^0 - \frac{K^2_{\kappa\tau}}{\epsilon^2_{\kappa \to \tau} - \epsilon^0} \tag{2.28}$$

$$E^2 = \epsilon^2 + \frac{K^2_{\kappa\tau}}{\epsilon^2_{\kappa \to \tau} - \epsilon^0} \tag{2.29}$$

The static DCI method can be illustrated by reference to the problem of the $2\pi + 2\pi$ cycloaddition of two ethylenes. The energies of the ground, singly excited and diexcited configurations are shown in Fig. 9. Using the notation of Figs. 1 and 2, we can symbolize the various configurations as follows:

$$\Psi^0_A = \psi^2_1 \psi^2_2 \tag{2.30}$$

$$\Psi^1_A = \psi^2_1 \psi^1_2 \psi^1_3 \tag{2.31}$$

$$\Psi^2_A = \psi^2_1 \psi^2_3 \tag{2.32}$$

$$\Psi^0_S = \phi^2_1 \phi^2_2 \tag{2.33}$$

$$\Psi^1_S = \phi^2_1 \phi^1_2 \phi^1_3 \tag{2.34}$$

$$\Psi^2_S = \phi^2_1 \phi^2_3 \tag{2.35}$$

The lower energy of Ψ^0_A relative to Ψ^0_S is a result we have obtained in a previous application of the one determinental OEMO method to the same problem. Now, CI will be greater in the case of the $2\pi_s + 2\pi_s$ complex. In most cases, A'_0 still ends up with a lower energy than S'_0 while S'_1 remains lower in energy than A'_1. Thus, the

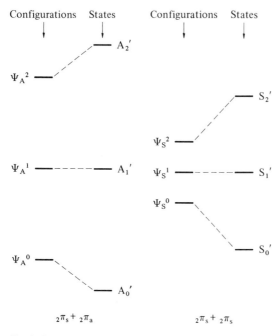

Fig. 9. Static DCI interaction diagram for $_2\pi_s + {_2}\pi_a$ and $_2\pi_s + {_2}\pi_s$ ethylene cyclodimerization

thermal reaction is expected to proceed in a $_2\pi_s + {_2}\pi_a$ sense and the photochemical reaction in a $_2\pi_s + {_2}\pi_s$ sense, *although there is nothing in this analysis which precludes that the opposite could happen.*

2.5 The Dynamic Delocalized Configuration Interaction Method

The dynamic DCI method is best exemplified by the Woodward-Hoffmann state correlation diagram approach. The necessary steps for implementing this approach are:

a) The fully delocalized MO's of the reaction complex **D–A** are constructed from the delocalized MO's of the fragments **D** and **A**. An approximate construction of this type can be accomplished by means of the OEMO method.

b) Ground, monoexcited, diexcited, etc., electronic configurations are built using the delocalized MO's of **D–A**.

c) The energy variation of the various configurations as a function of intermolecular distance is assessed by reference to the appropriate MO correlation diagram.

d) The configuration correlation diagram is constructed and the corresponding state correlation diagram can be deduced by identifying the configurations which interact with respect to the assumed Hamiltonian operator.

The $2\pi + 2\pi$ cycloaddition of two ethylenes can be used to illustrate this approach. The appropriate correlation diagrams are shown in Fig. 10.

The major drawback of the state correlation diagram constructed by using the dynamic DCI approach lies in its failure to show explicitly how barriers and intermediates arise on the ground and excited surfaces of certain classes of reactions. This point can be best illustrated by means of chemical examples.

First, in the cycloaddition of two ethylenes, proceeding in a $_2\pi_s + _2\pi_s$ manner, the state correlation diagram of Fig. 10a leads to the prediction of a barrier for the thermal reaction. However, there are three possible mechanisms which one can envision for the corresponding photochemical reaction initiated by $\pi\pi^*$ excitation of one ethylene partner. These three possibilities are shown in Fig. 11. Accordingly, no mechanistic inferences are possible. In fact, Woodward and Hoffmann chose case (b) of Fig. 11 as the representation of the mechanism of the photochemical $2\pi + 2\pi$ cycloaddition of two ethylenes and formulated their rules accordingly. However, we shall see that the actual mechanism of this and many other photoreactions is that implied by case (a) of Fig. 11.

Secondly, in the cycloaddition of two ethylenes proceeding via a $_2\pi_s + _2\pi_a$ path, the state correlation diagram suggests no barrier for the thermal reaction. Also, it

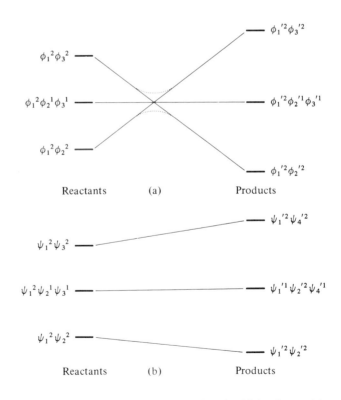

Fig. 10. State correlation diagrams for $_2\pi_s + _2\pi_s$ (a) and $_2\pi_s + _2\pi_a$ (b) cycloaddition. In case (a), the avoided crossing due to explicit account of two electron effects is indicated by dotted lines

29

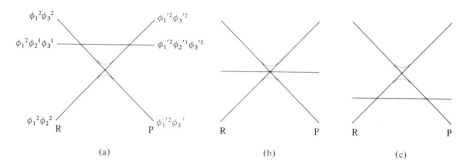

Fig. 11. Three possible qualitative state correlation diagrams for singlet $2\pi_s + 2\pi_s$ cycloaddition. R = reactants, P = product

predicts no barrier or excited intermediate formation for the photochemical reaction. However, chemical evidence strongly suggests that barriers are always present in thermal intermolecular cycloadditions and that their photochemical counterparts involve, in many cases, formation of excited intermediates.

An improvement of the state correlation diagram approach can be effected by rejecting the OEMO method for evaluating the energy of each configuration as a function of intermolecular distance. Once this attitude is adopted, state correlation diagrams can only be constructed with the aid of explicit calculations or physical-chemical intuition. In the former case, the simplicity and elegance of the method are lost. The intuitive approach is preferable but situations may arise where intuition is either inadequate or misleading.

3. The Dynamic Linear Combination
of Fragment Configurations Method

3.1 Definitions

In this book, we shall attempt to construct P.E. surfaces for representative organic reactions. These surfaces can provide direct information about many aspects of chemical reactivity.

Attention is focused primarily on the following reaction features:

a) Stereochemistry.
b) Regiochemistry.
c) Topochemistry.
d) The effect of substituents on reaction rate.
e) The selectivity-polarity relationship.
f) The possible existence of observable or isolable intermediates.

Accordingly, it is essential that we define the meaning of the above terms.

a) Stereochemistry. This term refers to the geometrical change within one or both reactants, such as bond rotation or inversion of configuration of an atomic center. We can distinguish three different types of reactions:

1. Reactions where both reactants undergo bond rotation or inversion. These will be designated T^+T^+ reactions.

2. Reactions where one reactant undergoes rotation or inversion but the other does not. These will be designated T^+T^- reactions.

3. Reactions where neither reactant undergoes rotation or inversion. These will be designated T^-T^- reactions.

The symbol T denotes geometrical change and the superscript the presence (positive sign) or absence (negative sign) thereof. In general, T^-T^- processes are examples of Least Motion (LM) stereochemical paths and T^+T^- processes are examples of Non Least Motion (NLM) stereochemical paths. The T^+T^+ processes are energetically unfavorable, at least in most cases of chemical interest, and will not be considered in this book.

Examples of T^-T^- reactions along with the labels to be used consistently in this work are shown below.

The Greek symbols specify the relevant orbital types and the numerical symbols signify the number of electrons involved in a chemical reorganization. The letter symbols denote the spatial relationship of reactant AO's which eventually give rise to bond formation, s standing for suprafacial and a for antarafacial. This nomenclature has been discussed by Woodward and Hoffmann [4a] and will not be pursued further here.

Examples of T^+T^- processes along with the appropriate labels are given below.

b) *Regiochemistry*. Once the manifold of P.E. surfaces for a given stereochemical path has been determined, one may identify the preferred regiochemistry associated with the particular stereochemical path. The term regiochemistry refers to the spatial orientation of two molecules or fragments which is maintained along the reaction coordinate and preserved in the final product. Typical examples of reactions which may yield regiochemically distinct products are the following:

32

Head to Head (HH) Head to Tail (HT)

Syn Anti

Exo Endo

Ortho Meta Para

c) *Topochemistry.* This term refers to the mode of association or dissociation of molecules or fragments when more than one possibility exists. A typical example of a reaction which can proceed via two topochemically distinct pathways is given below:

d) *The effect of substituents on reaction rates* can be expressed in a variety of ways. Firstly, substituents may alter drastically the ionization potential and electron affinity of a parent molecule and, as a result, the polarity of a corresponding reaction. This may produce a fundamental change in the shapes of the P.E. surfaces describing the reaction of the molecule in question. Secondly, substituents may be arranged in a symmetrical or unsymmetrical fashion within a given molecule and the imposed symmetry or lack thereof may determine whether two diabatic P.E. surfaces will cross or avoid each other. Thirdly, substituents control the "density" of the various P.E. surfaces. In future discussions, we shall be concerned with the above three effects of substituents on reaction rates, namely, the effect of polarity, the effect of unsymmetrical substitution and the effect of conjugative substitution, respectively.

e) Selectivity-polarity relationship. The term selectivity refers to the stereochemical preference (stereoselectivity), regiochemical preference (regioselectivity) or topochemical preference (toposelectivity) exhibited by a reaction of two molecules, **A** and **B**. As reaction polarity increases, the selectivity of a reaction may, in principle, decrease, remain unaltered, or, increase.

f) The LCFC approach, which constitutes the theoretical formalism to be utilized in the rest of this work, provides the basis for classifying reaction intermediates according to the type of wavefunction which describes them. We distinguish the following types of reaction intermediates:

1. M and M* intermediates. These arise from the interaction of nondegenerate locally excited, e.g., DA*, and charge transfer, e.g., D^+A^-, configurations. M intermediates are those in which the charge transfer configurations make the dominant contribution. M* intermediates are those in which the locally excited configurations make the dominant contribution.

2. M′ structure and M′* intermediate. These arise from the interaction of degenerate locally excited, e.g., DA*, and charge transfer, e.g., D^+A^-, configurations. The M′ structure constitutes the lower state and the M′* intermediate the upper state.

3. N and N* intermediates. These arise from the interaction of nondegenerate no bond, e.g., DA, and charge transfer, e.g., D^+A^-, configurations. N intermediates are those involving major no bond contribution and N* intermediates are those involving major charge transfer contribution.

4. N′ structure and N′* intermediate. These arise from the interaction of degenerate no bond, e.g., DA, and charge transfer, e.g., D^+A^-, configurations. The N′ structure is the lower state and the N′* intermediate is the upper state.

5. Ξ′ structure and Ξ′* intermediate. These arise from the interaction of degenerate no bond, e.g., DA, and diexcited, e.g., D*A*, configurations. The Ξ′ structure is the lower state and the Ξ′* intermediate is the upper state.

6. O and O* intermediates. These arise from the interaction of nondegenerate charge transfer, e.g. D^+A^-, and diexcited, e.g., D*A*, configurations. O intermediates are those in which the charge transfer configurations make the dominant contribution and O* intermediates are those where the reverse situation obtains.

At this point, an additional inclusive term will be introduced. Specifically, we define *chorochemistry* (Gr. *choros* = space) to be a term which collectively describes stereochemistry and regiochemistry. Finally, the following organic reactions classification scheme will be used in all subsequent discussions:

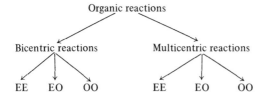

As we have already discussed, a multicentric reaction involves union of two atoms of one reactant with one or two atoms of a second reactant and orbital overlap between the uniting centers is maintained along the entire reaction coordinate. A bicentric reaction involves union of one atom of one reactant with one atom of a second reactant. The EE, EO, and OO reactions are defined in the following manner:

a) Even-even (EE) reactions. In this case, both ground state reactants have an even number of electrons.

b) Even-odd (EO) reactions. In this case, one ground state reactant has an even and the other an odd number of electrons.

c) Odd-Odd (OO) reactions. In this case, both ground state reactants have an odd number of electrons.

3.2 General Theoretical Considerations

Like the static LCFC approach, the dynamic LCFC approach is also based on the ZIDMOO approximation. However, unlike the situation which obtains in the static model, the energies of the configurations as well as the sizes of one electron interaction matrix elements, which depend on spatial orbital overlap, are functions of the intermolecular distance between donor and acceptor. Accordingly, the general procedure which is followed in developing a dynamic LCFC analysis of a chemical reactivity problem is the following:

a) The energies of the basis set configurations are plotted as a function of intermolecular distance. This amounts to the construction of P.E. *diabatic* surfaces, i.e., surfaces which obtain in the hypothetical case of zero CI.

The equations of the diabatic surfaces can be obtained by adding to each approximate static LCFC diagonal matrix element a quantity designated the steric function, S. This describes two effects:

1. The effect of steric repulsion, R, which is represented at the level of the ZIDMOO approximation by a sum of V_{ne}, V_{ee}, and V_{nn} terms, where V_{ne} is the nucleus-electron attraction, V_{ee} the electron-electron repulsion, and V_{nn} the nucleus-nucleus repulsion.

2. The effect of bond readjustment necessary for transforming reactants to products. In this case, the energy required for stretching or compressing bonds, denoted by L, is a function of the MO occupancy of the two reactants in the zero order basis set configurations. Thus, L is large for the DA diabatic surface because each reactant has a maximum bonding occupancy, i.e., the bonding MO's of each reactant are all occupied and any geometrical change which will tend to deemphasize the bonding AO interactions will raise the energy of the system significantly. In the case of the D^+A^- and DA* diabatic surfaces, L becomes smaller due to excitation of an electron from a bonding to an antibonding MO, i.e., a geometric deformation which will tend to deemphasize the bonding AO interactions will have a lesser effect on the energy of

35

the D^+A^- and DA* than on the energy of the DA diabatic surface. These conclusions are pertinent to closed shell reactants.

On the basis of the above, we distinguish two diabatic surface types:

1. *A diabatic surface of the no bond or local excitation type is repulsive.* The rise of the energy of such a diabatic surface as intermolecular distance decreases is described by the appropriate steric function.

2. *A diabatic surface of the charge transfer type has an attractive component and a repulsive component,* i.e., such a surface displays a minimum.

The steric functions can be defined empirically in a way which can lead to a quantitative estimate of reaction barriers. However, since we are interested in qualitative trends, we shall not define the various steric functions explicitly but, rather, restrict ourselves to the qualitative drawing of the shapes of the diabatic surfaces.

As an illustrative application, we consider the interaction of two closed shell molecules, **D** and **A**, leading to ionic product formation. We assume that the DA, D^+A^-, and DA* are the three lowest energy configurations sufficient for an adequate description of the reaction. A sketch of the diabatic surfaces is shown in Fig. 12. Note that the D^+A^- surface may or may not cross the DA surface, depending upon the particular case. In our example, a crossing occurs and the final product has ionic character. Under the assumption that the $HO^D - LU^A$ matrix element is zero and the $HO^D - HO^A$ one is non-zero, we are led to the construction of the final

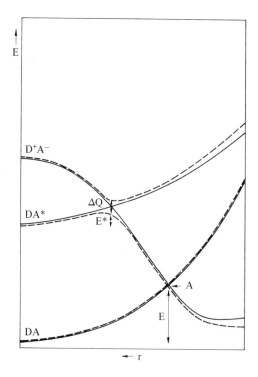

Fig. 12. Diabatic (solid lines) and adiabatic (dashed lines) surfaces for a hypothetical reaction of **D** and **A**

adiabatic surfaces as indicated in Fig. 12. It should be pointed out that depending upon the approximation employed in a theoretical analysis, two types of surface crossings can be differentiated:

a) Real crossings which occur irrespective of whether the interaction Hamiltonian contains only one electron or both one electron and two electron parts.

b) Pseudocrossings which occur when the interaction Hamiltonian contains only one electron parts.

By following the same line of reasoning we can differentiate between two types of avoided adiabatic surface crossings:

a) First degree avoided crossing due to the one electron mixing of the diabatic surfaces. This situation arises when the interaction Hamiltonian contains only one electron parts.

b) Second degree avoided crossing due to one electron plus two electron mixing of the diabatic surfaces. This situation arises when the interaction Hamiltonian contains one electron and two electron parts.

In dealing with pseudocrossings, the reader should remember that these are artifacts of the approximation involved in the dynamic LCFC method. Avoided crossings may be smaller or larger than actually depicted.

Let us now return to the adiabatic P.E. surfaces of Fig. 12. In the case of the thermal reaction, the reactants have to traverse a barrier in order to be converted to products. The height of the barrier is merely E. On the other hand, the photochemical reaction involves two distinct steps. First, a photochemical barrier E* should be surmounted. Second, the two reactants will find their way to the ground surface via a "hole" at A. This is a general pattern: in thermal reactions we shall be concerned only with barriers, while in photochemical reactions we have to worry about barriers *plus* decay processes.

While most chemists are familiar with the effect of barrier height upon reaction rate, the factors which control the efficiency of a decay process are many and, indeed, complicated. Various theoretical attempts towards understanding radiationless conversions have been made and quite a measure of success has been attained [25]. In order to develop an operationally workable scheme, we shall assume that the major factor which determines whether a given decay process will be fast is the energy gap ΔQ between the two surfaces involved. Specifically, we shall assume that as ΔQ decreases the decay efficiency increases. Other factors are also important but will be neglected in the formulation of general predictive rules.

3.3 The Interaction of Diabatic Surfaces

In the process of constructing adiabatic P.E. surfaces, we shall encounter two general diabatic surface interrelationships, namely, diabatic surfaces which cross and diabatic surfaces which do not do so. The interaction of such diabatic surfaces at the crossing point and/or away from it has important chemical consequences. In most cases, this interaction leads to formation of a barrier on the lower energy adiabatic surface and a potential energy well on the higher energy adiabatic surface. The energy gap, ΔQ, between the barrier maximum, which corresponds to the transition state of an elementary reaction step, and the minimum of the potential energy well, which corresponds to an excited intermediate, determines the efficiency of decay from the upper to the lower adiabatic surface. How do barrier heights, stabilities of intermediates, and decay efficiencies vary as the relative energies of the interacting diabatic surfaces change in a gradual manner? The most important diabatic surface energy variations crucial to this work are the following:

a) As reaction polarity increases, all diabatic surfaces which involve charge transfer from the donor to the acceptor are translated downwards in energy. Conversely, those which involve charge transfer in a reverse manner are translated upwards in energy.

b) As the excitation energy of the donor and/or the acceptor decreases, the corresponding locally excited diabatic surfaces are translated downwards.

The first problem concerns the energy lowering of a diabatic surface Ψ_1 due to its interaction with Ψ_2 and Ψ'_2, i.e., two pairwise interactions must be compared. Ψ_1 may or may not cross Ψ_2 and Ψ'_2. Hence, these two cases should be discussed separately.

We first focus attention on the case where Ψ_1, Ψ_2, and Ψ'_2 do not cross. At moderately long intermolecular distances, where spatial orbital overlap is appreciable, the difference between the energy gap of Ψ_1 and Ψ_2 and that of Ψ_1 and Ψ'_2, $\Delta\delta\epsilon$, may be larger, equal, or smaller in absolute magnitude than the difference between $\langle\Psi_1|\hat{P}|\Psi_2\rangle$ and $\langle\Psi_1|\hat{P}|\Psi'_2\rangle$, ΔP. In order to aid the presentation, we define the following quantities:

$$\epsilon(\Psi_1) - \epsilon(\Psi_2) = \delta\epsilon_{12} \tag{3.1}$$

$$\epsilon(\Psi_1) - \epsilon(\Psi'_2) = \delta\epsilon'_{12} \tag{3.2}$$

$$\Delta\delta\epsilon = |\delta\epsilon'_{12}| - |\delta\epsilon_{12}| \tag{3.3}$$

$$\langle\Psi_1|\hat{P}|\Psi_2\rangle = P_{12} \tag{3.4}$$

$$\langle\Psi_1|\hat{P}|\Psi'_2\rangle = P'_{12} \tag{3.5}$$

$$\Delta P = |P_{12}| - |P'_{12}| \tag{3.6}$$

We now distinguish the following categories:

a) If $\Delta\delta\epsilon$ and ΔP are both positive, the interaction of Ψ_1 and Ψ_2 will be stronger than that of Ψ_1 and Ψ'_2. This situation is illustrated in Fig. 13a.

b) If $\Delta\delta\epsilon$ is positive and ΔP zero, or, $\Delta\delta\epsilon$ zero and ΔP positive, the interaction of Ψ_1 and Ψ_2 will again be stronger. These situations are illustrated in Figs. 13b and 13c.

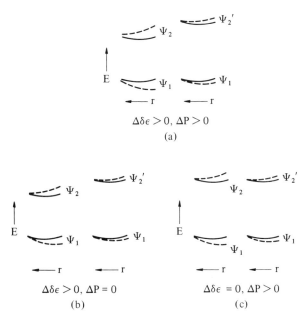

$\Delta\delta\epsilon > 0, \Delta P > 0$

(a)

$\Delta\delta\epsilon > 0, \Delta P = 0$ $\qquad\qquad$ $\Delta\delta\epsilon = 0, \Delta P > 0$

(b) $\qquad\qquad\qquad\qquad$ (c)

Fig. 13. Different patterns of noncrossing diabatic surface interactions. Diabatic surfaces are indicated by solid lines and adiabatic surfaces by dashed lines

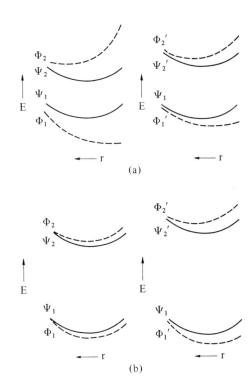

(a)

Fig. 14. a) Energy gap control of diabatic surface interaction. b) Matrix element control of diabatic surface interaction. In both cases, $\Delta\delta\epsilon > 0$ and $\Delta P < 0$. Solid lines indicate diabatic and dashed lines adiabatic surfaces

(b)

39

c) If $\Delta\delta\epsilon$ is positive and ΔP negative, the result will depend on the exact magnitude of the two quantities. If $|\Delta\delta\epsilon|$ exceeds $|\Delta P|$, the question of energy gap control or matrix element control may arise [21]. On the other hand, if $|\Delta P|$ exceeds $|\Delta\delta\epsilon|$, the interaction of Ψ_1 and Ψ_2' will become stronger.

Most of the problems we shall encounter fall within category (c) and involve $|\Delta\delta\epsilon| > |\Delta P|$. We can distinguish the two extreme situations illustrated in Fig. 14. In the first case, the Ψ_1 and the Ψ_2 and Ψ_2' diabatic surfaces are closely spaced in energy. Due to a smaller energy gap, the interaction of Ψ_1 and Ψ_2 is greater than that of Ψ_1 and Ψ_2'. This represents energy gap control of diabatic surface interaction which

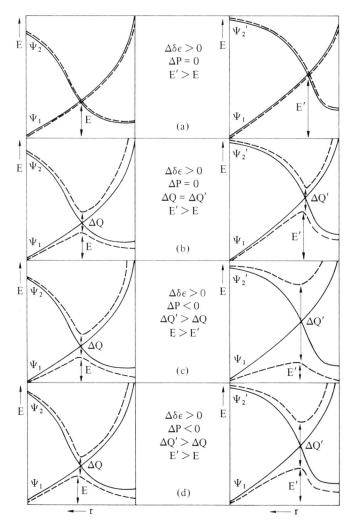

Fig. 15. Patterns of crossing diabatic surface interaction. Solid lines indicate diabatic and dashed lines adiabatic surfaces. Ψ_2 is displaced downwards in energy relative to Ψ_2'

causes the Φ_1 adiabatic surface to attain a lower energy than the Φ_1' adiabatic surface. The MO interaction analog has been discussed elsewhere [21]. In the second case, the Ψ_1 surface lies much below the Ψ_2 and Ψ_2' surfaces. Due to the more negative P_{12}', the interaction of Ψ_1 and Ψ_2' can now become stronger than that of Ψ_1 and Ψ_2. This represents matrix element control of diabatic surface interaction and can lead to a result opposite to that in the previous case. Once again, a discussion of the MO analog can be found in one of our recent works [21].

Consider next the case of three diabatic surfaces Ψ_1, Ψ_2, and Ψ_2', with Ψ_1 crossed by Ψ_2 and Ψ_2' at long intermolecular distances. We distinguish the following categories, where the symbols $\Delta\delta\epsilon$ and ΔP have the meaning specified previously:

a) $\Delta\delta\epsilon$ is positive and ΔP zero (Figs. 15a and 15b).

b) $\Delta\delta\epsilon$ and ΔP have opposite sign (Figs. 15c and 15d).

The most commonly encountered situations are those depicted by Fig. 15d.

We discussed above how the energy of a fixed diabatic surface Ψ_1 changes when a second diabatic surface Ψ_2, with which it can interact, is translated with respect to Ψ_1. We now consider the complementary problem of the energy change of a diabatic surface Ψ_2 relative to a fixed diabatic surface Ψ_1. We first consider cases where no diabatic surface crossing is involved. By reference to Fig. 16, we distinguish two situations:

a) The initial lowering of Ψ_2 relative to Ψ_2' (translation) is not overcompensated by the greater interaction of Ψ_2 and Ψ_1 and the adiabatic surface Φ_2 finds itself below the reference Φ_2' adiabatic surface. The translation effect dominates the interaction effect in this instance.

b) The initial lowering of Ψ_2 relative to Ψ_2' (translation) is overcompensated by the greater interaction of Ψ_2 and Ψ_1. The interaction effect dominates here.

A similar problem arises when Ψ_1 is translated relative to Ψ_2 and it can be analyzed along the same lines. In all subsequent discussions, we shall assume that a large translation is not counteracted by interaction.

A translation of a diabatic surface is the theoretical expression of a change in the electronic nature of the reactants. For example, a change of the electronic nature of the reactants leads to a change of reaction polarity which is theoretically expressed by a translation of charge transfer type diabatic surfaces. How do the interactions of diabatic surfaces control barrier heights and decay efficiencies?

The first example involves the interaction of two crossing diabatic surfaces Ψ_1 and Ψ_2'. A downward translation of Ψ_2' will lead to an earlier crossing and will have the following chemical consequences:

a) If the interaction matrix element is zero, the barrier E will decrease as the translation increases, while the energy gap between the resulting adiabatic surfaces, ΔQ, will be zero throughout (Fig. 15a).

b) If the interaction matrix element is nonzero, reduced spatial overlap will diminish the absolute magnitude of the interaction matrix element. As a result, the barrier will decrease if the translation is much greater than the change of the interaction matrix

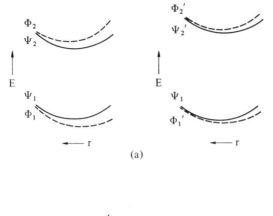

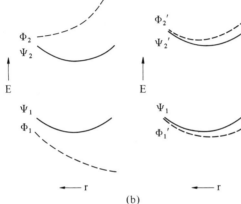

Fig. 16. The effect of translation of a diabatic surface Ψ_2' to Ψ_2 (solid lines) on the energy of the final corresponding adiabatic surface (dashed lines). a) Translation effect is dominant. b) Interaction effect is dominant

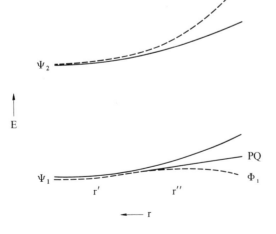

Fig. 17. Barrier formation via the interaction of two diabatic surfaces Ψ_1 and Ψ_2 away from a crossing point. The stabilization of Ψ_1 at r'' or the slope of PQ are indices of the height of the barrier and the looseness of the corresponding transition state. Solid lines indicate diabatic and dashed lines adiabatic surfaces

42

element (Fig. 15d) and vice versa (Fig. 15c). On the other hand, the energy gap between the resulting adiabatic surfaces, ΔQ, will shrink as translation increases in either case.

These ideas will be applied to the problem of the effect of polarity or excitation energy on barrier heights and decay efficiencies.

What happens if the interaction matrix element can take two different nonzero values? The difference between the corresponding interaction matrix elements is a function of overlap and will decrease as translation increases. As a result, the difference between the corresponding reaction barriers as well as the difference between the corresponding energy gaps will shrink. These ideas will be applied to the problem of the effect of polarity or excitation energy on choroselectivity.

Next, we consider the interaction of two diabatic surfaces, Ψ_1 and Ψ_2, away from an intended crossing point, if any exists (Fig. 17). As intermolecular distance decreases, the diabatic surfaces Ψ_1 and Ψ_2 begin to interact to an increasing extent thus forming two adiabatic surfaces. As downward translation of Ψ_2 increases and assuming that Ψ_1 and Ψ_2 interact strongly, a lower and earlier barrier is formed on the Φ_1 adiabatic surface. This is important and requires some illumination.

When the diabatic surfaces Ψ_1 and Ψ_2 interact at a large intermolecular distance, r', their interaction is near zero because the corresponding matrix element is very small due to negligible spatial overlap. At a moderate intermolecular distance, r'', interaction becomes possible and the energy of Ψ_1 is lowered while that of Ψ_2 is raised by the same amount, if overlap is neglected. The slope of the line PQ is an index of how fast the stabilization of Ψ_1 due to its interaction with Ψ_2 will tend to overtake the inherent energy increase of Ψ_1. In short, the stabilization of Ψ_1 at an arbitrary intermolecular distance can be given the meaning of a slope: as *stabilization increases due to a downward translation of Ψ_2, the barrier will tend to diminish and occur earlier on the reaction coordinate.* Once again, these ideas are pertinent to the problem of the effect of polarity or excitation energy on barrier heights and decay efficiencies.

Let us now consider what happens when the interaction matrix element can take two different values. Once again, the difference in the stabilization of Ψ_1 at an arbitrary intermolecular distance will have the meaning of a difference of two slopes. *As a downward translation of Ψ_2 occurs, the difference between the stabilization energies, and, hence, the corresponding slopes, will increase.*

These ideas are pertinent to the problem of the effect of polarity or excitation energy on selectivity.

3.4 Polarity Control of Barrier Heights and Decay Efficiencies

The effect of substituents upon the rate of a chemical reaction can be simply understood in terms of the ideas presented in the previous section. As an example, we shall consider once again the hypothetical reaction of closed shell **D** and **A** described by the three diabatic surfaces DA, D^+A^-, and DA*, and inquire as to how the rate and selectivity of the thermal reaction depend upon polarity (see Fig. 12).

The barrier of the thermal reaction of **D** and **A** arises as a result of the interaction of the DA and D^+A^- diabatic surfaces. An increase in polarity amounts to a downward translation of the D^+A^- diabatic surface. The consequences of this effect have been discussed in the previous section. It is predicted that as reaction polarity increases, the barrier of the thermal reaction of **D** and **A** decreases. Exceptions will arise if the change of polarity is outweighed by a change of the corresponding interaction matrix element.

Assuming that **D** and **A** can unite in two different ways, the energy difference of the corresponding thermal barriers provides a measure of the selectivity of the thermal reaction. Once again, an increase in polarity amounts to a downward translation of the D^+A^- diabatic surface. This leads to a decrease of selectivity.

The effect of polarity on the height of the photochemical barrier arising from the interaction of the DA* and D^+A^- diabatic surfaces as well as on the associated selectivity can be analyzed in a similar manner.

Finally, it is predicted that decay efficiency will increase as polarity increases. The reader should be able to analyze the reaction of **D** and **A** described by the three diabatic surfaces DA, D^+A^- and DA*, where DA does not cross D^+A^-.

3.5 The Effect of Excitation Energy on Photochemical Barrier Heights

If the D^+A^- diabatic surface is translated downwards in energy and the lowest locally excited diabatic surface, e.g., DA*, remains unaltered, the height of the photochemical barrier decreases, i.e. an increase in polarity is expected to be accompanied by a reduction of the barrier separating an encounter type complex and an excited intermediate,

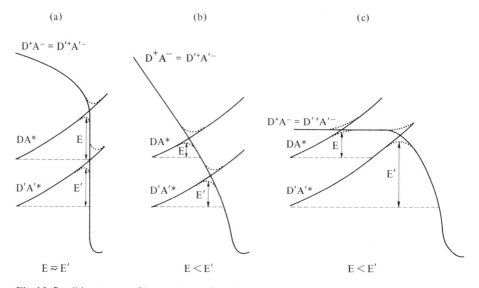

Fig. 18. Possible patterns of intersections of locally excited by charge transfer diabatic surfaces

everything else being constant. We next examine the complementary problem of the effect of an upward translation of the lowest locally excited diabatic surface, e.g., DA*, while D^+A^- remains unaltered.

Consider the interaction of **D** and **A** as well as the interaction of **D'** and **A'** such that $\epsilon(D^+A^-) = \epsilon(D'^+A'^-)$ and $\epsilon(DA^*) > \epsilon(D'A'^*)$, i.e., polarity effects are comparable but the excitation energies are different. Assuming that the translation of the locally excited surface dominates the counteracting effect of interaction at the crossing point, increased excitation energy will lead to a lower photochemical barrier, everything else being equal. This is illustrated in Fig. 18. In one extreme case, the DA* and D'A'* curves are crossed by the $D^+A^- = D'^+A'^-$ curve in the manner shown in Fig. 18c. In this event, $E' > E$. At the other extreme, the crossing is as shown in Fig. 18a. In such an event, $E' \simeq E$. The situation actually encountered is of the intermediate type illustrated in Fig. 18b. In such an event, we expect that $E' > E$. In short, the photochemical barrier decreases as excitation energy increases, everything else being constant.

3.6 Diabatic Surface Interrelationships. A Classification of Chemical Reactions

One operationally useful classification of chemical reactions is according to the interrelationship of the DA and D^+A^- diabatic surfaces. We distinguish the following types of reactions:

a) *Nonionic reactions,* where the DA diabatic surface does not cross the D^+A^- diabatic surface.

b) *Ionic reactions,* where DA crosses D^+A^- but the minimum of the D^+A^- diabatic surface lies above the minimum of the DA diabatic surface.

c) *Superionic reactions,* where DA crosses D^+A^- and the minimum of the D^+A^- diabatic surface lies below the minimum of the DA diabatic surface.

A second useful classification is according to the interrelationship of the D^+A^- and DA* diabatic surfaces, where DA* is assumed to be the lowest energy locally excited diabatic surface. Specifically, we can distinguish the following possible situations:

a) The DA* and D^+A^- diabatic surfaces do not cross, D^+A^- lies always above DA* and their energy separation is very large (P1' photoreaction). In this case, the photochemical reaction barrier will be very large and no excited intermediate will be formed because the two diabatic surfaces can interact only to a very small extent. Hence, a P1' photoreaction will be expected to be very inefficient. This situation is illustrated in Fig. 19a.

b) The DA* and D^+A^- diabatic surfaces do not cross, D^+A^- lies always above DA*, and their energy separation is small (P1'' photoreaction). In this case, the photochemical barrier will become smaller and an excited M* intermediate could be formed because the two diabatic surfaces can interact strongly. Hence, a P1″ can be more efficient than a P1′ photoreaction. This situation is illustrated in Fig. 19b.

c) The DA* and D^+A^- diabatic surfaces cross and the minimum of the D^+A^- lies above the energy of DA* at infinite intermolecular distance (P2′ photoreaction). In this case, the photochemical barrier will be small, an excited M intermediate will be formed, and the reaction can be highly efficient *(vide infra)*. This situation is illustrated in Fig. 19c.

d) The DA* and D^+A^- diabatic surfaces cross and the minimum of the D^+A^- curve lies below the energy of the DA* at infinite intermolecular distance (P2″ photoreac-

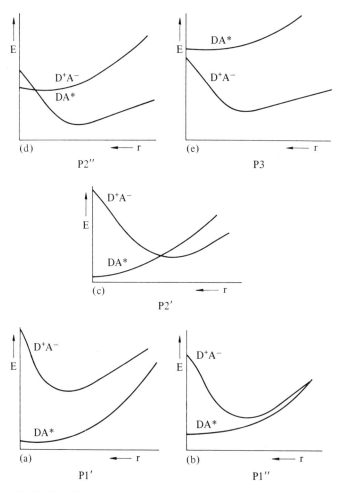

Fig. 19. Classification of nonionic chemical reactions according to the interrelationship of the lowest energy excited and charge transfer diabatic surfaces. In this example, DA* is the lowest excited diabatic surface

tion). In this case, the photochemical barrier will be small, an excited M intermediate will be formed having greater stability than any **D** . . . **A*** encounter complex, and the reaction can be highly efficient *(vide infra)*. This situation is illustrated in Fig. 19 d.

e) The DA* and D^+A^- diabatic surfaces do not cross and D^+A^- lies always below DA* (P3 photoreaction). This case is expected to occur very rarely and will not be dealt with any further (Fig. 19 e).

The above classification is not restricted to photoreactions initiated by $\pi\pi^*$ photo-excitation. Thus, similar considerations apply to the case of $n\pi^*$ photoexcitation. Furthermore, the DA* and D^+A^- diabatic surfaces may be both singlets or both triplets, i.e. the classification scheme is applicable to singlet as well as triplet photoreactions.

The above discussion clearly suggests that, in photochemical reactions where the energy of the DA* surface remains relatively fixed and only the D^+A^- surface varies appreciably in energy, the order of photoreaction efficiency will be $P2'' > P2' > P1''$ $> P1'$. *This is a prediction which will be valid only if decay of the excited intermediate M* or M does not lead to energy wastage and photoreactivity is primarily controlled by the height of the photochemical barrier.* Energy wastage is a process which is not expected for most P1' or P1'' reactions, but may become a realistic possibility for many P2'' reactions. In general, radiationless decay of M to ground state reactants will become significant when the energy gap between the DA diabatic surface and the minimum of the D^+A^- diabatic surface becomes small. When this condition is met, energy wastage may upset the predicted trend. As we shall see, this state of affairs is sometimes encountered in photochemical reactions. Conversely, if we assume that for a series of photochemical reactions the energy of the D^+A^- surface remains relatively fixed and only the energy of the DA* (or D*A) surface varies appreciably in energy, it follows that the height of the photochemical barrier will decrease as the energy of DA* (or D*A) increases. Accordingly, photoreaction efficiency will increase in the order $P2'' > P2' > P1'' > P1'$ under the assumptions stated before, i.e., no appreciable energy wastage and barrier control of photoreactivity.

Once a reaction has been classified, prediction of various reactivity aspects can be made. Clearly, what an experimentalist needs is some simple way of identifying the type of reaction at hand in order to anticipate its features. In proposing such an identification procedure, approximations and generalizations are unavoidable. The following scheme constitutes only a model classification of chemical reactions which will undoubtedly be improved and refined further.

The proposed definitions are as follows:

a) Nonionic reaction: $\quad I_D - A_A - C_m > 1.5$ eV

b) Ionic reactions: $\quad\quad I_D - A_A - C_m < 1.5$ eV

c) P1' reactions: $\quad\quad\ \ I_D - A_A - C_m - G(\pi\pi^*) > 3$ eV

d) P1'' reaction: $\quad\quad\ 3$ eV $> I_D - A_A - C_m - G(\pi\pi^*) > 2$ eV

e) P2' reaction: $\quad\quad\ \ 2$ eV $> I_D - A_A - C_m - G(\pi\pi^*) > 0$ eV

f) P2'' reaction: $\quad\quad\ \ 0 > I_D - A_A - C_m - G(\pi\pi^*)$

In the above inequalities, C_m is an average coulombic attractive term for the range of intermolecular distances between 2.5 and 3.0 Å and the quantity $I_D - A_A - C_m$ is assumed to be a satisfactory approximation of the energy of the D^+A^- diabatic surface near its minimum. Furthermore, $G(\pi\pi^*)$ is the energy of DA^* (or D^*A) at infinite intermolecular distance. Each definition is proposed on the basis of explicit test calculations and considerations of the experimental data. For example, the condition for an ionic reaction is that $I_D - A_A - C_m$ should be less than 1.5 eV because the barrier of most thermal ionic reactions is of the order of 1 eV.

The inequalities given above can be evaluated by reference to experimental or calculated ionization potentials, electron affinities, and spectroscopic transitions. The C_m term can be easily calculated; however, in order to facilitate the procedure, the following C_m values are recommended:

a) In reactions of simple molecules which are not highly conjugated, C_m will be taken equal to 5 eV.

b) In reactions of highly conjugated molecules, C_m will be taken equal to 4 eV.

c) In reactions where the positive hole is localized away from the framework where the negative excess is delocalized, C_m will be taken equal to 3 eV.

Typical examples of these three cases are:

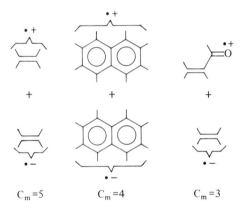

$C_m = 5$ $C_m = 4$ $C_m = 3$

While the distinction between nonionic and ionic reactions by reference to the diabatic surface interrelationships is always unambiguous, reactions can be envisioned which appear to be ionic although the diabatic surface interrelationship would indicate otherwise. These reactions can be termed *pseudoionic* and they can materialize under the following circumstances:

a) An increase in solvent polarity translates the adiabatic surface containing primarily a D^+A^- contribution downwards in energy so that it intersects the originally lower lying adiabatic surface containing primarily a DA contribution. These considerations are illustrated schematically in Fig. 20.

b) The D^+A^- diabatic surface minimum lies above the DA diabatic surface but the interaction of D^+A^- with higher lying diabatic surfaces causes the diabatic surface arising from D^+A^- to intersect the adiabatic surface arising from DA. This presupposes that DA and D^+A^- do not interact to any appreciable extent. Pseudoionicity arises

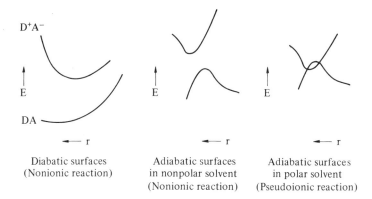

Diabatic surfaces	Adiabatic surfaces	Adiabatic surfaces
(Nonionic reaction)	in nonpolar solvent	in polar solvent
	(Nonionic reaction)	(Pseudoionic reaction)

Fig. 20. Solvent induced pseudoionic reaction

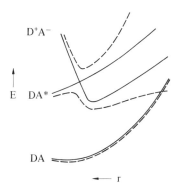

Fig. 21. Intrinsic pseudoionic reaction

frequently in highly polar nonionic reactions. These considerations are illustrated schematically in Fig. 21.

In the following chapters, we shall draw P.E. surfaces and discuss their chemical implications. As our approach will be comparative, it will be necessary to assume that certain reaction variables remain relatively constant. For emphasis, we state the assumptions which are implicit in our analysis of chemical reactivity:

a) In comparing two different chorochemical or topochemical modes of union of two reactants, it will be assumed that steric repulsions are comparable. Any differential steric destabilization will be superimposed upon the result of the analysis based on this assumption. Thus, for example, the P.E. surfaces for $_2\pi_s + _2\pi_s$ and $_2\pi_s + _2\pi_a$ cycloadditions will be drawn by assuming that steric effects are comparable. The fact that the latter path is sterically destabilized relative to the former will be kept in mind in reaching a final conclusion.

b) In comparing two different chorochemical or topochemical modes of union of two reactants, it will be assumed that attractive electrostatic effects, entering via the C term in the equation for the energy of a charge transfer diabatic surface, are com-

49

parable if the two modes of union involve similar accumulation of atomic centers within a unit volume. For example, electrostatic effects will be assumed comparable in the case of $_2\pi_s + _2\pi_s$ and $_2\pi_s + _2\pi_a$ cycloadditions. By contrast, electrostatic effects will not be comparable in exo vs. endo $_4\pi_s + _2\pi_s$ cycloaddition, i.e., the endo form will be clearly favored.

c) In drawing qualitative P.E. surfaces, no attempt is made to depict the thermodynamic relationship of reactants and products. Presently, this cannot be done in any simple qualitative manner. Rather, we trace the energy of the reaction system from some long intermolecular distance to some short intermolecular distance which is still longer than that of the product. As a result, *thermodynamic effects on reaction rates should be considered separately from our approach.*

d) The decay efficiency enters as an important parameter in the overall rate expression for a photochemical reaction in the sense that unfavorable decay implies a long lived excited intermediate which can be intercepted by solute and/or solvent or revert back to reactants.

e) In discussing the effect of polarity on the rate of a reaction, we assume that inter-action matrix elements remain *relatively* constant, unless otherwise stated. We caution that the validity of this assumption should be further studied.

4. Even-Even Intermolecular Multicentric Reactions

4.1 Potential Energy Surfaces for $2\pi + 2\pi$ Cycloadditions

In this section, we shall develop the P.E. surfaces for $\pi + \pi$ cycloadditions. The model reaction is depicted below and *it will be assumed that the two ethylenes approach each other in a symmetrical fashion.* We shall restrict our attention to singlet reactivity.

$$2 \; = \!\!\! = \quad \longrightarrow \quad \square$$

Fig. 22 specifies the basis set of singlet zero order configurations to be employed in the analysis. Symmetry adaptation is required since the two reaction partners are identical and, thus, charge displacement and localization of excitation are unacceptable. The various (symmetry adapted) configurations are:

$$\Psi_1 = D_1 D_2 \tag{4.1}$$

$$\Psi_2^+ = \frac{1}{\sqrt{2}} \, (D_1^+ D_2^- + D_1^- D_2^+) \tag{4.2}$$

$$\Psi_2^- = \frac{1}{\sqrt{2}} \, (D_1^+ D_2^- - D_1^- D_2^+) \tag{4.3}$$

$$\Psi_3^+ = \frac{1}{\sqrt{2}} \, (D_1^* D_2 + D_1 D_2^*) \tag{4.4}$$

$$\Psi_3^- = \frac{1}{\sqrt{2}} \, (D_1^* D_2 - D_1 D_2^*) \tag{4.5}$$

$$\Psi_4^+ = \frac{1}{\sqrt{2}} \, (D_1^{+2} D_2^{-2} + D_1^{-2} D_2^{+2}) \tag{4.6}$$

$$\Psi_4^- = \frac{1}{\sqrt{2}} \, (D_1^{+2} D_2^{-2} - D_1^{-2} D_2^{+2}) \tag{4.7}$$

$$\Psi_5^+ = \frac{1}{\sqrt{2}} \; (D_1^{**}D_2 + D_1 D_2^{**}) \tag{4.8}$$

$$\Psi_5^- = \frac{1}{\sqrt{2}} \; (D_1^{**}D_2 - D_1 D_2^{**}) \tag{4.9}$$

$$\Psi_6^+ = \frac{1}{\sqrt{2}} \; (D_1^{+*}D_2^- + D_1^- D_2^{+*}) \tag{4.10}$$

$$\Psi_6^- = \frac{1}{\sqrt{2}} \; (D_1^{+*}D_2^- - D_1^- D_2^{+*}) \tag{4.11}$$

$$\Psi_7^+ = \frac{1}{\sqrt{2}} \; (D_1^+ D_2^{-*} + D_1^{-*} D_2^+) \tag{4.12}$$

$$\Psi_7^- = \frac{1}{\sqrt{2}} \; (D_1^+ D_2^{-*} - D_1^{-*} D_2^+) \tag{4.13}$$

$$\Psi_8 = D_1^* D_2^* \tag{4.14[1]}$$

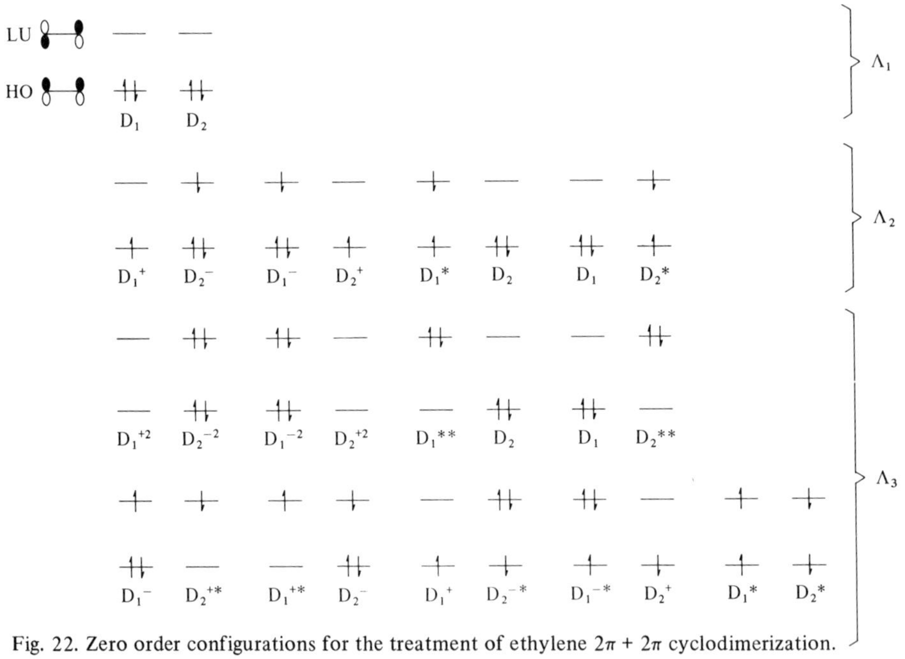

Fig. 22. Zero order configurations for the treatment of ethylene $2\pi + 2\pi$ cyclodimerization. Λ_1, Λ_2, and Λ_3 are packets

---

[1] In the case of four open shell electrons, there exist two singlet, three triplet, and one quintuplet states. For simplicity, we have included in our basis set only the lower energy singlet. This can be viewed as a combination of two fragments each in its lowest triplet state.

52

While in the static LCFC method the number of basis set configurations was restricted to only five, a more extended basis set is needed in the case of the dynamic LCFC method. In the former case, the properties of a reaction complex at a relatively large intermolecular distance were investigated. By contrast, in the dynamic LCFC method the evolution of the reaction complex is followed along the entire reaction coordinate.

The various zero order configurations fall within three sets. One set includes only the no bond configuration, DA. The second set includes configurations which arise by promoting one electron to a higher energy MO. Henceforth, such configurations will be termed collectively monoexcited configurations. The third set includes configurations which arise by promoting two electrons to higher energy MO's. Henceforth, such configurations will be termed collectively diexcited configurations. These three sets define three *packets*, Λ_1, Λ_2 and Λ_3. A similar classification is possible for the (symmetry adapted) configurations. The three packets behave differently depending upon the nature of the stereochemical pathway.

The various no bond, monoexcited and diexcited configurations can be subdivided into smaller categories as follows:

a) No Bond Configuration.   DA   No Bond

b) Monoexcited Configurations. $\left\{ \begin{array}{l} D^+A^-, D^-A^+ \\ DA^*, D^*A \end{array} \right.$ Charge Transfer
Locally Excited

c) Diexcited Configurations. $\left\{ \begin{array}{l} DA^{**}, D^{**}A \\ D^{+2}A^{-2}, D^{-2}A^{+2} \\ D^{+*}A^-, D^-A^{+*}, \\ D^+A^{-*}, D^{-*}A^+ \\ D^*A^* \end{array} \right.$ Locally Diexcited
Charge Ditransfer
$\left. \begin{array}{c} \\ \\ \end{array} \right\}$ Charge Excited
Triplet Diexcited

We next consider the relative energies of the configurations as a function of intermolecular distance. Recalling our assumption that the zero order basis set configurations can be mixed only in a one electron sense, we can write:

$$\epsilon(\Psi_1) \simeq S \tag{4.15}$$

$$\epsilon(\Psi_2^+) = \epsilon(\Psi_2^-) = \epsilon(D_1^+ D_2^-) \simeq I - A + C + S' \tag{4.16}$$

$$\epsilon(\Psi_3^+) = \epsilon(\Psi_3^-) = \epsilon(D_1^* D_2) \simeq {}^1G + S' \tag{4.17}$$

$$\epsilon(\Psi_4^+) = \epsilon(\Psi_4^-) = \epsilon(D_1^{+2} D_2^{-2}) \simeq 2(I - A) + C' + S'' \tag{4.18}$$

$$\epsilon(\Psi_5^+) = \epsilon(\Psi_5^-) = \epsilon(D_1^{**} D_2) \simeq 2\,{}^1G + S'' \tag{4.19}$$

$$\epsilon(\Psi_6^+) = \epsilon(\Psi_6^-) = \epsilon(D_1^{+*} D_2^-) = \epsilon(\Psi_7^+) = \epsilon(\Psi_7^-) = \epsilon(D_1^+ D_2^{-*}) \tag{4.20}$$
$$\simeq (I - A) + {}^1G + C'' + S''$$

$$\epsilon(\Psi_8) \simeq 2\,{}^3G + S'' \tag{4.21}$$

53

In the above equations, the various symbols have the following meaning:

^{1}G: Singlet $\pi\pi^*$ excitation energy of ethylene.

^{3}G: Triplet $\pi\pi^*$ excitation energy of ethylene.

I: Ionization potential of ethylene.

A: Electron affinity of ethylene.

C, C', C'': Coulombic terms.

Particular note should be made of the quantity ^{3}G which represents the $\pi\pi^*$ triplet excitation energy of ethylene since Ψ_8 is taken to be the spin adapted configuration which arises from two $\pi\pi^*$ triplet ethylenes. It should be emphasized that these equations are empirical and very approximate in nature, but they are expected to provide a realistic picture of the relative energy ordering of the various basis configurations at a given intermolecular distance.

The steric function is a composite of two terms, one representing intermolecular nonbonded repulsion and another representing the energy necessary for bond readjustment. The former term, symbolized by R, depends on the bulk of the two reactants. The latter term, symbolized by L, depends on the diabatic surface type. The variation of L in the case at hand is easy to understand. Specifically, in the process

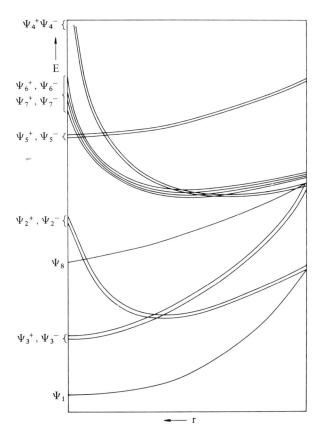

Fig. 23. Diabatic energy surfaces for the treatment of ethylene $2\pi + 2\pi$ cyclodimerization

of forming cyclobutane, the C=C ethylenic bond stretches. The energy requirement for this bond readjustment will decrease as the number of bonding pi electrons of the two reactants decreases. In the no bond configuration there are four bonding pi electrons and the energy required for bond readjustment will be maximal. In the monoexcited configurations, there are two net bonding electrons and the required energy will be smaller. Finally, in the diexcited configurations, there are no net bonding electrons and the energy demand will be minimal. This analysis indicates that the repulsive part of a diabatic surface may be steepest for the no bond configuration, i.e., $S > S' > S''$.

A plot of the energies of the basis set configurations as a function of intermolecular distance gives rise to diabatic surfaces (Fig. 23). In this respect, equations 4.15–4.21 are nothing else but the equations of these diabatic surfaces which can be grouped into three distinct *diabatic surface packets*, Λ_1, Λ_2, and Λ_3. Reactivity trends will depend on the way in which the diabatic surfaces within each packet interact with each other as well as the way in which diabatic surfaces of one packet interact with those of another packet.

How do the interactions of the various diabatic surfaces depend upon the nature of the stereochemical path? The interaction matrix is shown in Table 2 and the intra- and interpacket interactions are depicted schematically below.

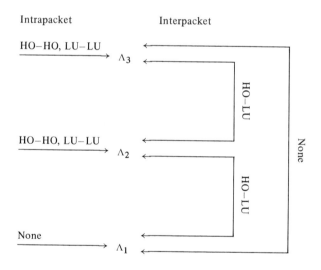

Λ_1: Ψ_1
Λ_2: Ψ_2^+, Ψ_2^-, Ψ_3^+, Ψ_3^-
Λ_3: Ψ_4^+, Ψ_4^-, Ψ_5^+, Ψ_5^-, Ψ_6^+, Ψ_6^-, Ψ_7^+, Ψ_7^-, Ψ_8

We first consider the $_2\pi_s + _2\pi_s$ geometry of approach and intrapacket interactions. Since the Λ_1 packet contains only Ψ_1, there are no such interactions to discuss. The Λ_2 packet contains four configurations and the only possible interactions involve Ψ_3^+ and Ψ_3^- with Ψ_2^+ and Ψ_2^-.

Table 2. Interaction matrix for $2\pi + 2\pi$ cyclodimerization of ethylene

	Ψ_1	Ψ_2^+	Ψ_2^-	Ψ_3^+	Ψ_3^-	Ψ_4^+	Ψ_4^-
Ψ_1		$(HO^{D1}-LU^{D2})$ 0 $+$ $(HO^{D2}-LU^{D1})$	0	0		0	0
Ψ_2^+			0	0	0	$(HO^{D1}-LU^{D2})$ 0 $+$ $(HO^{D2}-LU^{D1})$	0
Ψ_2^-				0	$(HO^{D1}-HO^{D2})$ 0 $+$ $(LU^{D1}-LU^{D2})$	0	$(HO^{D1}-LU^{D2})$ $+$ $(HO^{D2}-LU^{D1})$
Ψ_3^+					0	0	0
Ψ_3^-						0	0
Ψ_4^+							0
Ψ_4^-							
Ψ_5^+							
Ψ_5^-							
Ψ_6^+							
Ψ_6^-							
Ψ_7^+							
Ψ_7^-							
Ψ_8							

The configurations of this packet can interact strongly in a HO–HO and/or LU–LU sense. Finally, the Λ_3 packet contains nine configurations which can interact strongly also in a HO–HO and/or LU–LU sense. It follows that diabatic surfaces within each packet interact strongly in the case of a $2\pi_s + 2\pi_s$ geometry of approach.

Let us now consider interpacket interactions for the same geometry. For example, consider the interaction of Ψ_1 belonging to Λ_1 with Ψ_2^+, Ψ_2^-, Ψ_3^+ and Ψ_3^- belonging to Λ_2.

Ψ_5^+	Ψ_5^-	Ψ_6^+	Ψ_6^-	Ψ_7^+	Ψ_7^-	Ψ_8
0	0	0	0	0	0	0
0	0	0	0	0	0	$(HO^{D2}-LU^{D1})$ + $(HO^{D1}-LU^{D2})$
0	0	0	0	0	0	0
0	0	$(HO^{D1}-LU^{D2})$ + $(HO^{D2}-LU^{D1})$	0	0	0	0
0	0	0	$(HO^{D1}-LU^{D2})$ + $(HO^{D2}-LU^{D1})$	0	$(HO^{D1}-LU^{D2})$ + $(HO^{D2}-LU^{D1})$	0
0	0	$LU^{D1}-LU^{D2}$	0	$HO^{D1}-HO^{D2}$	0	0
0	0	0	$LU^{D1}-LU^{D2}$	0	$HO^{D1}-HO^{D2}$	0
0		$(LU^{D1}-LU^{D2})$ + $(HO^{D1}-HO^{D2})$	0	$(LU^{D1}-LU^{D2})$ + $(HO^{D1}-HO^{D2})$	0	0
		$(LU^{D1}-LU^{D2})$ + $(HO^{D1}-HO^{D2})$	0	$(LU^{D1}-LU^{D2})$ + $(HO^{D1}-HO^{D2})$	0	0
			0	0	0	$(HO^{D1}-HO^{D2})$ + $(LU^{D1}-LU^{D2})$
				0	0	0
				0	0	$(HO^{D1}-HO^{D2})$ + $(LU^{D1}-LU^{D2})$
						0

The possible interaction matrix elements will be zero, i.e., the HO–LU overlap integral is zero. The same situation obtains in the case of interaction between diabatic surfaces of Λ_2 and those of Λ_3. It follows that diabatic surfaces of different packets do not interact in the $2\pi_s + 2\pi_s$ geometry of approach.

How does product formation occur in a $2\pi_s + 2\pi_s$ cycloaddition of two ethylenes? As the two ethylenes approach each other, spatial overlap increases and intrapacket interactions continuously increase while interpacket interactions always remain zero.

As a result, the Λ_2 and Λ_3 packets "swell" and their boundaries travel toward the Ψ_1 diabatic surface. Eventually, the Λ_3 packet will define the product potential well and Λ_2 will define the potential well where an excited intermediate will be accommodated. The resulting P.E. surfaces which describe the photochemical and thermal $_2\pi_s + _2\pi_s$ cycloadditions can be represented as shown in Fig. 24. The curve depicted by a solid line represents the Ψ_1 diabatic surface which cannot interact with any other diabatic surface in a one electron sense. Secondly, the curve depicted by dots represents the boundary of the Λ_2 packet which is generated by the interaction of the Λ_2 diabatic surfaces. Finally, the curve depicted by a dashed line represents the boundary of the Λ_3 packet which is generated by the interaction of the Λ_3 diabatic surfaces.

We now consider in some detail the features of the ground and lowest excited surfaces of $_2\pi_s + _2\pi_s$ cycloaddition which are crucial for the understanding of the kinetics of the thermal and photochemical reactions. The ground surface displays a barrier, E, which is produced by the pseudocrossing of the Λ_1 and Λ_3 packets. The key features of the excited surface can be understood by examining the intrapacket interactions of Λ_2. The avoided crossing of the Ψ_2^- and Ψ_3^- diabatic surfaces leads to formation of a barrier, E^*, and an excited intermediate, M. If the interaction of Ψ_2^- and Ψ_3^- is very strong, the barrier will tend to disappear. Past the minimum describing the excited intermediate, M, Ψ_2^- crosses the boundary of Λ_3, and this is a real crossing. At this point, a small geometrical distortion of the reaction complex removes the symmetry prohibition of $\Lambda_2 - \Lambda_3$ interaction. As a result, the reaction

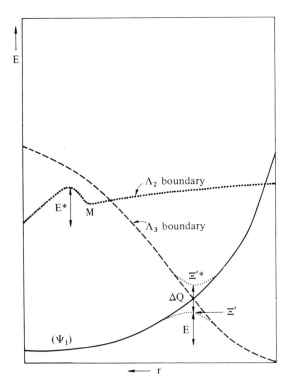

Fig. 24. Potential energy surfaces for thermal and photochemical $_2\pi_s + _2\pi_s$ cycloaddition of two ethylenes. The $\Psi_1 - \Lambda_3$ boundary avoided crossing due to electron interaction is indicated by dotted lines. The reaction is actually pseudoionic

system makes the transition to the boundary of Λ_3 and is ultimately led to become an excited intermediate Ξ'^*. This can now decay to the ground surface structure Ξ' and the efficiency of the decay will depend on the energy gap, ΔQ, between the ground and lowest excited surfaces. The magnitude of the energy gap is controlled by the two electron interaction of the Λ_1 and Λ_3 packets. Finally, Ξ' can be transformed to products or reactants.

The various features of the P.E. surfaces can be conveyed also by means of chemical equations[1].

a) Thermal $_2\pi_s + {}_2\pi_s$.

$$\mathbf{D}_1 + \mathbf{D}_2 \rightarrow (\mathbf{D}_1 \ldots \mathbf{D}_2) \rightarrow C_4H_8$$

b) Photochemical $_2\pi_s + {}_2\pi_s$.

$$\mathbf{D}_1 + \mathbf{D}_2^* \rightarrow (\mathbf{D}_1 \ldots \mathbf{D}_2^*) \rightarrow M \rightarrow \Xi'^* \begin{cases} \rightarrow \mathbf{D}_1 + \mathbf{D}_2 \\ \rightsquigarrow C_4H_8 \end{cases}$$

with $\mathbf{D}_1 + \mathbf{D}_2$ above and $\mathbf{D}_1 + \mathbf{D}_2 + h\nu$ below M.

In the above equations, $\mathbf{D}_1 \ldots \mathbf{D}_2$ is a thermal encounter complex which can be weakly bound (N type complex) and $\mathbf{D}_1 \ldots \mathbf{D}_2^*$ is an excited encounter complex which also can be weakly bound (M* type complex)[2].

In the above discussion, it was suggested that the Λ_3 packet was responsible for product formation and the Λ_2 packet for excited intermediate formation. In the $_2\pi_s + {}_2\pi_s$ cyclodimerization of ethylene, and, in fact, in $_2\pi_s + {}_2\pi_s$ cycloadditions, in general, simple symmetry considerations can be employed to rationalize these assignments. Specifically, each zero order configuration, i.e., DA, D^+A^-, D^-A^+, etc., can be labeled as symmetric or antisymmetric by reference to the MO electron occupancy in each case. The symmetry of a given reactant MO is defined with respect to a symmetry element of the individual reactant which is preserved during reaction. As an example, we consider the symmetry designation of the DA zero order configuration in the $_2\pi_s + {}_2\pi_s$ cycloaddition of two ethylenes. The appropriate symmetry element is the mirror plane which passes through the midpoint of the C=C bond. The DA configuration has S symmetry because all four pi electrons occupy the bonding ethylenic MO's which have S symmetry, i.e., a more explicit symmetry designation is SSSS.

By following similar reasoning, it can be shown that the zero order configurations of Λ_1 and Λ_3 have S symmetry while those of Λ_2 have A symmetry. Since the final

---

1 In all chemical equations to be encountered henceforth, a wavy arrow indicates radiationless decay. Furthermore, no attempt is made to indicate reversibility.

2 The possible existence of such weak complexes can be justified on the basis of the rigorous LCFC treatment. By contrast, the LCFC-ZIDMOO approach overestimates their stability. For further discussion see Chap. 22.

product is a closed shell molecule having S symmetry, it follows that the Λ_3 packet will give rise to product formation, the Λ_2 packet will generate the excited intermediate and, finally, the Λ_1 packet will create a high lying excited state of the product.

An obvious advantage in using the symmetry designations of the zero order configurations concerns the ready evaluation of interaction matrix elements, i.e., two zero order configurations having the same symmetry can interact. The classification of the various zero order configurations according to symmetry appropriate to a $2\pi_s + 2\pi_s$ interaction of the two ethylenes is given in Table 3. In addition, symmetry arguments can be used to identify the interacting configurations, i.e., Ψ_1, Ψ_2^+, Ψ_2^-, etc.

We now are prepared to formulate certain useful rules. Thus, $2\pi_s + 2\pi_s$ cyclo-additions, in general, exhibit the following features:

a) The barrier of the thermal reaction is determined by the pseudocrossing of Λ_1 and Λ_3.

b) The barrier of the photochemical reaction is determined by intrapacket interactions of Λ_2 and the decay process by the pseudocrossing of Λ_1 and Λ_3.

The key aspects of such reactions can be determined by considering only the following surfaces:

a) The no bond diabatic surface.

b) The lowest energy charge resonance and excitation resonance diabatic surfaces, or, the lowest charge transfer and locally excited diabatic surfaces when nonidentical reactants are involved.

Table 3. Symmetry designation of zero order basis set configurations for $2\pi_s + 2\pi_s$ and $2\pi_s + 2\pi_a$ cyclodimerizations

Packet	Zero order configuration	$2\pi_s + 2\pi_s$ Symmetry designation	$2\pi_s + 2\pi_a$ Symmetry designation
Λ_1	$D_1 D_2$	SSSS $=$ S	SSAA $=$ S
Λ_2	$D_1^+ D_2^-$	SSSA $=$ A	SSAA $=$ S
	$D_1^- D_1^+$	SSSA $=$ A	SSAA $=$ S
	$D_1^* D_2$	SSSA $=$ A	SSSA $=$ A
	$D_1 D_2^*$	SSSA $=$ A	AAAS $=$ A
Λ_3	$D_1^{++} D_2^{=}$	SSAA $=$ S	SSAA $=$ S
	$D_1^{=} D_2^{++}$	SSAA $=$ S	SSAA $=$ S
	$D_1^{**} D_2$	SSAA $=$ S	SSSS $=$ S
	$D_1 D_2^{**}$	SSAA $=$ S	AAAA $=$ S
	$D_1^{+*} D_2^-$	SSAA $=$ S	SSSA $=$ A
	$D_1^- D_2^{+*}$	SSAA $=$ S	AAAS $=$ A
	$D_1^{-*} D_2^+$	SSAA $=$ S	SSSA $=$ A
	$D_1^+ D_2^{-*}$	SSAA $=$ S	AAAS $=$ A
	$D_1^* D_2^*$	SSAA $=$ S	SSAA $=$ S

c) The boundary of the Λ_3 packet responsible for final product formation.

We next consider the $_2\pi_s + {}_2\pi_a$ geometry of approach. Once again, the Λ_1 packet contains only Ψ_1 so there are no intrapacket interactions. The Λ_2 packet contains four configurations and all the possible one electron interaction matrix elements are of the HO–HO or LU–LU type and, hence, zero. The same situation exists in the case of intrapacket interactions in Λ_3. Contrary to the case of $_2\pi_s + {}_2\pi_s$ cycloaddition, the diabatic surfaces within each packet do not interact when the two ethylenes approach each other in a $_2\pi_s + {}_2\pi_a$ manner.

The interpacket interactions also differ from those which obtain in the $_2\pi_s + {}_2\pi_s$ geometry. For example, consider the interaction of Ψ_1 of Λ_1 with Ψ_2^+, Ψ_2^-, Ψ_3^+ and Ψ_3^- of Λ_2. The possible interaction matrix elements will be large, i.e., the HO–LU overlap integral is large. The same situation obtains in the case of interaction between diabatic surfaces of Λ_2 and those of Λ_3. In short, diabatic surfaces of different packets do interact in the $_2\pi_s + {}_2\pi_a$ geometry of approach. Once again, the situation contrasts with that encountered in the $_2\pi_s + {}_2\pi_s$ cycloaddition.

How does product formation occur in a $_2\pi_s + {}_2\pi_a$ cycloaddition of two ethylenes? As the two molecules approach each other spatial overlap and interpacket interactions progressively increase while intrapacket interactions always remain zero. As a result, the three packets "repel" each other. The P.E. surfaces for photochemical and thermal $_2\pi_s + {}_2\pi_a$ cycloadditions are shown in Fig. 25. The curves in solid lines represent the adiabatic surfaces which arise from the interaction of Ψ_1, Ψ_2^+ and Ψ_8. The curves in dotted lines are the adiabatic surfaces which arise from the original diabatic Ψ_3^+ and Ψ_3^- surfaces which cannot interact with any of the diabatic Ψ_1, Ψ_2^+, Ψ_2^- and Ψ_8 surfaces. Finally, the curve in dashed line is the adiabatic surface which arises from the original diabatic Ψ_2^- surface which cannot interact with any of the Ψ_1, Ψ_2^+, Ψ_3^+, Ψ_3^- and Ψ_8 diabatic surfaces.

The P.E. surface aspects of the $_2\pi_s + {}_2\pi_a$ cycloaddition are simple to understand. The barrier of the thermal reactions, E, arises primarily from the interaction of the Ψ_1 and Ψ_2^+ diabatic surfaces. The photochemical barrier, E*, arises from the crossing of the charge transfer and locally excited diabatic surfaces; its height is controlled by the interaction of the Ψ_1, Ψ_2^+ and Ψ_8 diabatic surfaces. Decay of the excited intermediate to the ground adiabatic surface completes the sequence of events in the photoreaction.

The details of the P.E. surfaces for $_2\pi_s + {}_2\pi_a$ cycloaddition can be expressed as follows:

a) Thermal $_2\pi_s + {}_2\pi_a$ cycloaddition.

$$D_1 + D_2 \rightleftarrows (D_1 \ldots D_2) \rightarrow C_4H_8$$

b) Photochemical $_2\pi_s + {}_2\pi_a$ cycloaddition.

$$D_1 + D_2^* \rightleftarrows (D_1 \ldots D_2^*) \rightarrow 0 \begin{array}{c} C_4H_8 \\ \uparrow \\ \text{(} \rightarrow C_4H_8 + h\nu \\ \text{)} \searrow D_1 + D_2 + h\nu' \\ \downarrow \\ D_1 + D_2 \end{array}$$

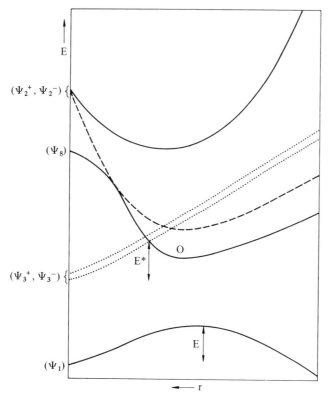

Fig. 25. Potential energy surfaces for thermal and photochemical $_2\pi_s + _2\pi_a$ cycloaddition of two ethylenes. The parent diabatic surfaces are indicated in parentheses

As illustrated before, symmetry considerations can be helpful in the evaluation of interaction matrix elements. The symmetry designations of the zero order basis set configurations for $_2\pi_s + _2\pi_a$ cyclodimerization of ethylene and, in fact, $_2\pi_s + _2\pi_a$ cycloadditions, in general, are shown in Table 3.

Once again, the above discussion leads to the formulation of simple rules. Thus, $_2\pi_s + _2\pi_a$ cycloadditions, in general, exhibit the following features:

a) The barrier of the thermal reaction is determined by the interaction of Λ_1 and Λ_2.

b) The barrier of the photochemical reaction is determined by the interpacket interactions of Λ_1, Λ_2 and Λ_3 and the decay process by the interaction of Λ_1 and Λ_2.

The key aspects of a reaction like a $_2\pi_s + _2\pi_a$ cycloaddition can be determined by considering only the following surfaces:

a) The no bond diabatic surface.

b) The charge resonance and excitation resonance diabatic surfaces, or, the lowest charge transfer and locally excited diabatic surfaces when nonidentical reactants are involved.

62

c) The lowest energy diexcited diabatic surface.

While we have restricted our attention to the $_2\pi_s + _2\pi_s$ and $_2\pi_s + _2\pi_a$ mechanisms of cyclobutane formation, a third stepwise mechanism exists as shown below:

The first step of this mechanism constitutes a bicentric reaction where the geometry of approach of the two ethylenes is the one indicated below. This geometry of approach is designated the Transoid Biradicaloid (TB) geometry.

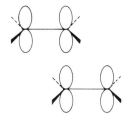

Henceforth, we shall refer to the above mechanism as the TB mechanism and the first elementary step as the TB step.

In bicentric reactions, both intra- and interpacket interactions obtain. Accordingly, the P.E. surfaces of the thermal and photochemical TB step constitute intermediate cases.

Having discussed the P.E. surfaces for $_2\pi_s + _2\pi_s$, TB and $_2\pi_s + _2\pi_a$ cycloadditions, the question arises as to which of the three stereochemical paths is favored. In the thermal reaction, only the height of the ground surface barrier has to be considered. While an *a priori* prediction cannot be made, it is reasonable to assume that interpacket interactions will produce the lowest barrier in the case of $_2\pi_s + _2\pi_a$ cycloaddition. The prediction of the relative efficiencies of the $_2\pi_s + _2\pi_s$, TB and $_2\pi_s + _2\pi_a$ photoreactions is more straightforward. In this instance, one has to be concerned with two aspects of the reaction, namely, the barrier on the lowest excited surface and the decay to ground surface. The $_2\pi_s + _2\pi_s$ stereochemical path is favored on both counts: the height of the photochemical barrier is optimum and the efficiency of decay maximum since the reaction system can find a "hole" for decay to the ground surface, i.e., it can arrive at a point (intermediate $\Xi'*$) where two adiabatic surfaces are separated by a small energy gap, and, hence, decay from the upper to lower surface is expected to be fast.

The above considerations lead naturally to the formulation of two rules:

a) The preferred stereochemical pathway of a thermal reaction will be the one which maximizes interpacket interactions. We shall refer to this rule as the Thermal Reactivity Rule (TRR).

b) The preferred stereochemical pathway of a photochemical reaction will be the one which maximizes intrapacket interactions and minimizes interpacket interactions.

A reaction path which is favorable by reference to the above rules will be termed aromatic. Paths which fullfill the contrary conditions will be termed antiaromatic.

4.2 The Effect of Polarity. Potential Energy Surfaces for Nonionic and Ionic $2\pi + 2\pi$ Cycloadditions

The $2\pi + 2\pi$ cycloaddition of two ethylenes is a reaction characterized by low polarity. How are the conclusions of the previous section modified when polarity is increased, i.e., when one ethylene is replaced by a donor olefin and the other by an acceptor olefin? The basis set configurations appropriate to this problem are shown in Fig. 26, the interaction matrix is given in Table 4, and the diabatic surfaces can be inferred from equations 4.15–4.21.

The effect of increasing polarity on the energies of the diabatic surfaces is simple to understand. Specifically, increasing polarity will depress the energies of D^+A^- and $D^{+2}A^{-2}$ type diabatic surfaces. This, in turn, will produce a depression of the energies

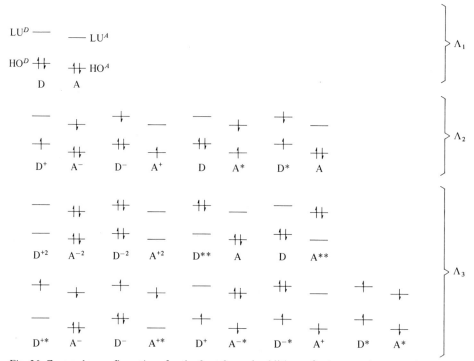

Fig. 26. Zero order configurations for the $2\pi + 2\pi$ cycloadditions of a donor and an acceptor molecule. Λ_1, Λ_2 and Λ_3 are packets

of the adiabatic surfaces which are generated from these diabatic surfaces. Obviously, increasing polarity will have the opposite effect on D^-A^+ and $D^{-2}A^{+2}$ diabatic surfaces. These considerations lead us to expect that a point will be reached where all successive packets will interpenetrate, e.g., DA, which belongs to Λ_1, will intersect D^+A^-, which belongs to Λ_2. Accordingly, the shapes of the adiabatic P.E. surfaces will be considerably altered. In this chapter, we shall be concerned exclusively with nonionic and ionic cycloadditions.

Consider the diabatic surfaces DA, D^+A^-, DA*, and the boundary of the Λ_3 packet which are sufficient to describe adequately a $_2\pi_s + _2\pi_s$ cycloaddition. An increase in reaction polarity leads to a lowering of the D^+A^- diabatic surface as well as a lowering of the boundary of the Λ_3 packet. The latter is an indirect effect which can be understood in simple terms. Specifically, Λ_3 contains charge excited diabatic surfaces which are depressed in energy as reaction polarity increases. As a result, these charge excited diabatic surfaces interact more strongly with the lowest energy diabatic surface of Λ_3, e.g., D*A*, and this leads to a depression of the Λ_3 boundary. Ultimately, D^+A^- will cross DA, marking the transition from a nonionic to an ionic cycloaddition. These con-

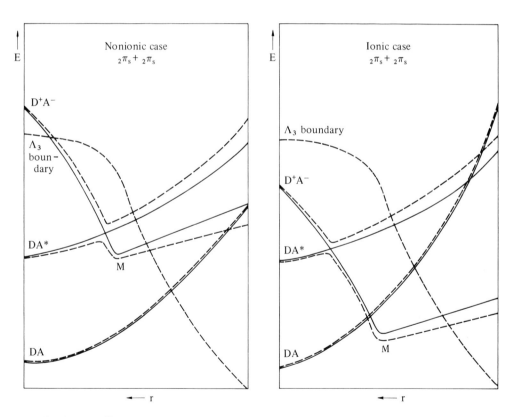

Fig. 27. The effect of polarity on the shapes of the potential energy surfaces for $_2\pi_s + _2\pi_s$ cycloaddition. Solid lines indicate diabatic and dashed lines adiabatic surfaces. Two electron interactions are neglected

Table 4. Interaction matrix for $2\pi + 2\pi$ cycloadditions

	DA	D^+A^-	D^-A^+	D^*A	DA^*	$D^{+2}A^{-2}$	$D^{-2}A^{+2}$
DA	–	HO^D-LU^A	HO^A-LU^D	0	0	0	0
D^+A^-		–	0	LU^D-LU^A	HO^D-HO^A	HO^D-LU^A	0
D^-A^+			–	HO^D-HO^A	LU^D-LU^A	0	HO^A-LU^D
D^*A				–	0	0	0
DA^*					–	0	0
$D^{+2}A^{-2}$						–	0
$D^{-2}A^{+2}$							–
$D^{**}A$							
DA^{**}							
$D^{*+}A^-$							
D^+A^{-*}							
D^-A^{+*}							
$D^{-*}A^+$							
D^*A^*							

siderations are illustrated in Fig. 27. *The excited intermediate M of a nonionic reaction becomes the ground intermediate M of an ionic reaction.*

Next, consider the diabatic surfaces, DA, D^+A^-, DA*, and D^*A^* which are sufficient to describe adequately a $_2\pi_s + _2\pi_a$ cycloaddition. As before, a progressive increase in reaction polarity leads to a progressive lowering of the D^+A^- diabatic surface. At some point D^+A^- crosses DA and an ionic $_2\pi_s + _2\pi_a$ cycloaddition materializes. This is illustrated in Fig. 28. The nature of the excited intermediate depends on the degree of interaction of the DA, D^+A^-, and D^*A^* diabatic surfaces. If D^+A^- interacts dominantly with DA, an N* intermediate is formed, while, if D^+A^- interacts dominantly with D^*A^*, an O intermediate materializes.

How does the rate of a cycloaddition depend on reaction polarity? The rate of a nonionic thermal $_2\pi_s + _2\pi_s$ cycloaddition can be discussed by reference to Fig. 27 and the minimal kinetic scheme shown below. As polarity increases,

$$D + A \xrightarrow{\text{ k }} \text{Product}$$

the boundary of the Λ_3 packet is translated downwards in energy intersecting the DA diabatic surface at longer intermolecular distances and resulting in a lower barrier and a larger k.

The rate of an ionic thermal $_2\pi_s + _2\pi_s$ cycloaddition can be discussed by reference to Fig. 27 and the minimal kinetic scheme shown below, where the experimental rate constant, k, is related to k_1, k_{-1}, and k_2 by the expression $k = \dfrac{k_2 K}{K[D]_0 + 1}$, where $K = k_1/k_{-1}$. As polarity increases, K increases. On the other hand, k_2 remains relatively unaffected because *both* the D^+A^- diabatic surface and the Λ_3 boundary are translated downwards in energy, albeit not to the same extent. Accordingly, it is predicted that an increase in polarity will result in a larger k.

$D^{**}A$	DA^{**}	$D^{+*}A^-$	D^+A^{-*}	D^-A^{+*}	$D^{-*}A^+$	D^*A^*
0	0	0	0	0	0	0
0	0	0	0	0	0	$HO^A - LU^D$
0	0	0	0	0	0	$HO^D - LU^A$
0	0	$HO^D - LU^A$	0	0	$HO^A - LU^D$	0
0	0	0	$HO^D - LU^A$	$HO^A - LU^D$	0	0
0	0	$LU^D - LU^A$	$HO^D - HO^A$	0	0	0
0	0	0	0	$LU^D - LU^A$	$HO^D - HO^A$	0
—	0	$LU^D - LU^A$	0	0	$HO^D - HO^A$	0
	—	0	0	0	0	$HO^D - HO^A$
		—	0	0	0	$LU^D - LU^A$
			—	0	0	$HO^D - HO^A$
				—	—	$LU^D - LU^A$
						—

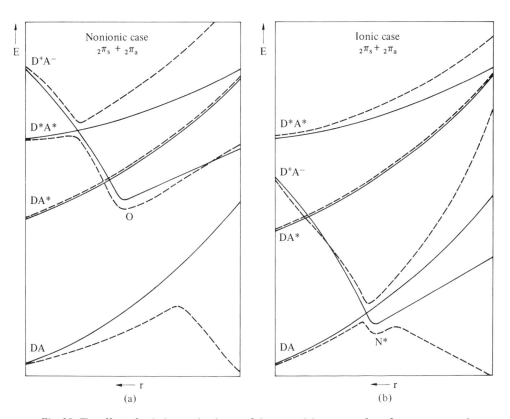

Fig. 28. The effect of polarity on the shapes of the potential energy surfaces for $2\pi_s + 2\pi_a$ cyclo-addition. Solid lines indicate diabatic and dashed lines adiabatic surfaces. Two electron interactions are neglected

$$D + A \underset{k_{-1}}{\overset{k_1}{\rightleftharpoons}} M \overset{k_2}{\longrightarrow} \text{Product}$$

The polarity dependence of the rate of a nonionic $_2\pi_s + {_2}\pi_s$ photocycloaddition can be discussed by reference to Fig. 27 and the minimal kinetic scheme shown below, where it is assumed that the rate of decay of M to ground state reactants is minimal while the radiationless decay of Ξ'^* to ground state reactants is very fast. It is further assumed that the partition of Ξ'^* to ground state reactants and products is polarity independent.

$$D + A^* \underset{k_{-1}}{\overset{k_1}{\rightleftharpoons}} M \overset{k_2}{\longrightarrow} \text{Product (Ground State)}$$

As polarity increases and the D^+A^- diabatic surface is translated downwards in energy, k_1/k_{-1} increases and k_2 remains relatively unaffected. Accordingly, we predict that, under that stated assumptions, an increase in polarity will lead to an increase of the rate of the $_2\pi_s + {_2}\pi_s$ photocycloaddition.

The rate of a nonionic thermal $_2\pi_s + {_2}\pi_a$ cycloaddition can be discussed by reference to Fig. 28 and the minimal kinetic scheme shown below.

$$D + A \overset{k}{\longrightarrow} \text{Product}$$

In most cases, an increase in reaction polarity will result in an increase of the rate of a nonionic $_2\pi_s + {_2}\pi_a$ cycloaddition. Exceptions will be found when small polarity differences are counteracted by large $\langle DA | \hat{P} | D^+A^- \rangle$ matrix element differences.

The rate of an ionic thermal $_2\pi_s + {_2}\pi_a$ cycloaddition can be discussed by reference to Fig. 28 and the minimal kinetic scheme shown below:

$$D + A \underset{k_{-1}}{\overset{k_1}{\rightleftharpoons}} N^* \overset{k_2}{\longrightarrow} \text{Product}$$

Assuming that formation of N^* is rate determining, we are led to the same conclusions as in the nonionic case.

The polarity dependence of the rate of a nonionic $_2\pi_s + {_2}\pi_a$ photocycloaddition can be discussed by reference to Fig. 28 and the simple kinetic scheme shown below, where it is assumed that the partition of O to ground state reactants and products is polarity independent.

$$D + A^* \underset{k_{-1}}{\overset{k_1}{\rightleftharpoons}} O \overset{k_d}{\longrightarrow} \text{Product (Ground State)}$$

In the above equation, k_d is the rate constant for decay of O to ground state products. An increase in polarity leads to an increase in k_1/k_{-1} and k_d and faster product formation.

A better understanding of the effect of polarity on barrier heights and intermediate stabilities can be achieved from a comparative study of the two configuration interaction patterns shown below.

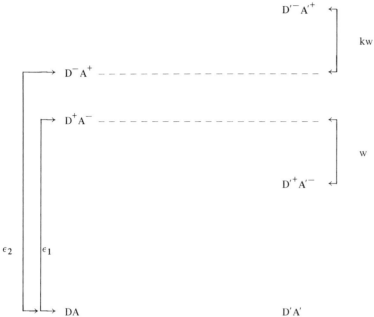

Assuming that all interaction matrix elements are equal to θ, the energy depressions of the no bond configurations due to interactions are given by the following expressions:

Pattern A:
$$\Delta E_1 = \frac{\theta^2}{\epsilon_1} + \frac{\theta^2}{\epsilon_2} \tag{4.22}$$

Pattern B:
$$\Delta E_2 = \frac{\theta^2}{\epsilon_1 - w} + \frac{\theta^2}{\epsilon_2 + kw} \tag{4.23}$$

The second equation defines a parabolic curve for the interval $\epsilon_1 > w > -\epsilon_2/k$. When $k = 1$, an increase of polarity, i.e., an increase of w, is certain to produce the result $\Delta E_2 > \Delta E_1$. However, when k exceeds unity, a small increase of polarity may result in an initial decrease of the energy depression of the no bond configuration due to interaction. The trend will be reversed as polarity continues to increase. In all subsequent discussions we shall assume that a *large* polarity increase, i.e., a large w, will be accompanied by an increase of ΔE. In chemical terms, this means increased stability of a molecule or lower barrier of a chemical reaction. One should always remain cognizant of the fact that the polarity effect, i.e., the energy variation of D^+A^-, can be counteracted by the variation of the interaction matrix elements in the two cases which are being compared.

The details of the adiabatic surfaces for nonionic and ionic $_2\pi_s + _2\pi_s$ and $_2\pi_s + _2\pi_s$ cycloadditions can be best conveyed by means of the following chemical equations:

Even-Even Intermolecular Multicentric Reactions

a) Nonionic thermal $_2\pi_s + _2\pi_s$ cycloaddition.

$$D + A \rightarrow (D \ldots A) \rightarrow P \quad (P = \text{Product})$$

b) Ionic thermal $_2\pi_s + _2\pi_s$ cycloaddition.

$$D + A \rightarrow (D \ldots A) \rightarrow M \rightarrow P$$

c) Nonionic photochemical $_2\pi_s + _2\pi_s$ cycloaddition.

d) Ionic photochemical $_2\pi_s + _2\pi_s$ cycloaddition*.

e) Nonionic thermal $_2\pi_s + _2\pi_a$ cycloaddition.

$$D + A \rightarrow (D \ldots A) \rightarrow P$$

f) Ionic thermal $_2\pi_s + _2\pi_a$ cycloaddition.

$$D + A \rightarrow (D \ldots A) \rightarrow N^* \rightarrow P$$

g) Nonionic photochemical $_2\pi_s + _2\pi_a$ cycloaddition

h) Ionic photochemical $_2\pi_s + _2\pi_a$ cycloaddition.

In conclusion, we wish to point out that the photochemical nonionic $2\pi + 2\pi$ cycloadditions treated in this chapter were assumed to be of the P2' or P2'' type and the photochemical ionic $2\pi + 2\pi$ cycloadditions of the P2'' type. Extension of our treatment to the other types of $2\pi + 2\pi$ photocycloadditions is trivial.

4.3 Pericyclic, Effectively Pericyclic and Quasipericyclic Reactions

An important aspect revealed by an adiabatic surface is reaction pericyclicity which is intimately connected with reaction stereoselectivity. In Figs. 29 and 30, we depict the parts of adiabatic surfaces which involve pericyclic bonding and those which do not. The following trends are noteworthy:

a) In the case of thermal nonionic $2\pi_s + 2\pi_s$ cycloaddition, the reaction system jumps from a nonpericyclic surface to a pericyclic surface in the neighborhood of A. As polarity increases, this transition occurs earlier in the reaction coordinate while simultaneously the height of the thermal barrier decreases.

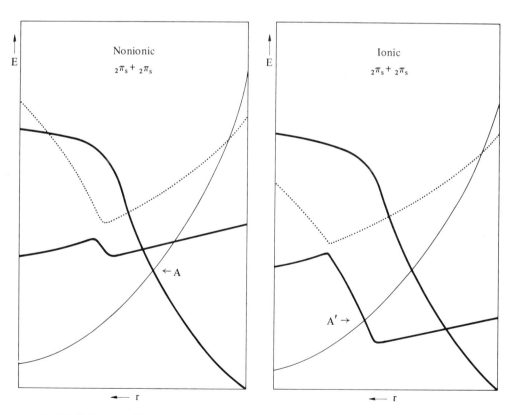

Fig. 29. Path pericyclicities in $2\pi_s + 2\pi_s$ cycloaddition (symmetrical complex).
—— = nonpericyclic path; ▬▬ = pericyclic path; ······ = antipericyclic path

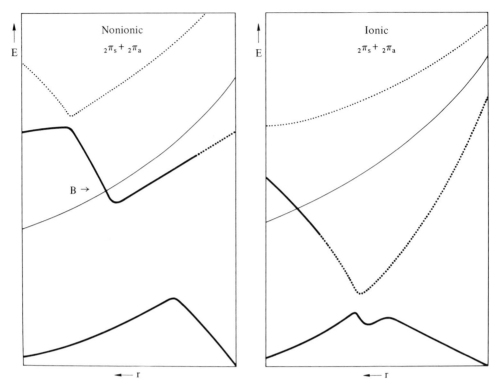

Fig. 30. Path pericyclicities in $_2\pi_s + _2\pi_a$ cycloaddition (symmetrical complex).
——— = nonpericyclic path;   ▬▬▬ = pericyclic path;   · · · · · · = antipericyclic path

b) In the case of thermal ionic $_2\pi_s + _2\pi_s$ cycloaddition, the reaction system jumps from a nonpericyclic surface to a pericyclic surface in the neighborhood of A', i.e, very early on the reaction coordinate.

c) In the case of photochemical nonionic or ionic $_2\pi_s + _2\pi_s$ cycloadditions, the reaction complex traverses a surface which is entirely pericyclic.

d) In the case of thermal nonionic or ionic $_2\pi_s + _2\pi_a$ cycloadditions, the reaction complex traverses a surface which is entirely pericyclic.

e) In the case of photochemical nonionic $_2\pi_s + _2\pi_a$ cycloadditions, *and assuming a dominant interaction of* D^+A^- *with* D^*A^*, the reaction system jumps from a nonpericyclic to a pericyclic surface in the neighborhood of B. This transition occurs very early on the reaction coordinate.

f) In the case of photochemical ionic $_2\pi_s + _2\pi_a$ cycloaddition, the reaction system jumps from a nonpericyclic surface onto a surface which is initially pericyclic and subsequently becomes antipericyclic.

The analysis presented above provides grounds for an important classification of thermal reactions. Thus, we distinguish the following reaction types:

a) A *pericyclic reaction* is one which proceeds entirely on a pericyclic path, i.e., a path which is defined by a sequence of points each corresponding to a pericyclic electronic state. Thermal nonionic and ionic $_2\pi_s + _2\pi_a$ cycloadditions are all pericyclic reactions.

b) An *effectively pericyclic reaction* is one which proceeds mostly, but not entirely on a pericyclic path. In such a reaction, the two reactants approach each other on a nonpericyclic path, i.e., a path defined by a sequence of points each corresponding to a nonpericyclic electronic state. When spatial overlap between the two reactants is still small, the reaction complex makes a transition to a pericyclic path. In general, this transition occurs near the top of a barrier on the ground surface. Thermal ionic $_2\pi_s + _2\pi_s$ cycloadditions are examples of effectively pericyclic reactions.

c) A *quasipericyclic reaction* is one which proceeds to a large extent on a non-pericyclic path. In such a reaction, the two reactants approach each other on a non-pericyclic path and a transition to a pericyclic path occurs late in the reaction when spatial overlap between the two reactants is large. Once more, this transition occurs near the top of a barrier on the ground surface. Thermal nonionic $_2\pi_s + _2\pi_s$ cycloadditions are examples of quasipericyclic reactions.

By following similar reasoning, one may classify photochemical reactions according to the nature of the path.

4.4 The Effect of Unsymmetrical Substitution and the Effect of Conjugative Substitution

The unsymmetrical and conjugative substitution effects differ from the normal polarity effect although the substituents which enforce the former may also necessarily enforce the latter effect.

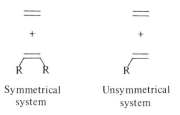

Symmetrical   Unsymmetrical
system   system

Consider the two $2\pi + 2\pi$ cycloadditions shown above. In these systems, intra-packet interactions ($HO^D - HO^A$ and $LU^D - LU^A$) are comparable for the $_2\pi_s + _2\pi_s$ mode and interpacket interactions ($HO^D - LU^A$ and $HO^A - LU^D$) are comparable for the $_2\pi_s + _2\pi_a$ and TB modes as indicated by comparing the forms of the appropriate interaction matrix elements for a fixed intermolecular distance. On the other hand, interpacket interactions for the $_2\pi_s + _2\pi_s$ mode and intrapacket interactions for the $_2\pi_s + _2\pi_a$ mode are nonzero in the unsymmetrical and zero in the symmetrical system. In short, unsymmetrical substitution has as a principal effect the "turning on" of inter- or intrapacket interactions which are absent in the symmetrical system. As a result, the barrier of the thermal $_2\pi_s + {}_2\pi_s$ cycloaddition and the barrier of the photochemical $_2\pi_s + {}_2\pi_a$ cycloaddition can be reduced drastically.

Unsymmetrical substitution may also have a pronounced effect on the nature of a given intermediate. For example, consider the thermal $_2\pi_s + {}_2\pi_s$ ionic cycloaddition. In symmetrical cases, D^+A^- can only mix with DA^* and D^*A and the intermediate is of the M type. In unsymmetrical cases, D^+A^- can mix in addition with DA and, if this latter mixing predominates, an N^* intermediate will materialize. This situation may arise when the reactants are highly unsymmetrical. In such an event, the N^* intermediate, in contrast to the M intermediate, will be devoid of pericyclic bonding.

We now turn our attention to a different problem and consider the two reaction systems shown below. In this case, one ethylene in system 2 is so substituted as to introduce a low lying D^*A^* diabatic surface. Since the thermal barrier of the $_2\pi_s + {}_2\pi_s$ cycloaddition is a function of ^3G_D plus ^3G_A, it follows that, as the triplet energies of the two reactants decrease, the thermal barrier for the $_2\pi_s + {}_2\pi_s$ will diminish. System 2 is more polar than system 1, and this may also contribute to a diminished $_2\pi_s + {}_2\pi_s$ thermal barrier.

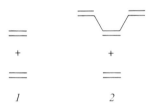

     $+$ $+$

     1   *2*

4.5 The Regiochemistry of $2\pi + 2\pi$ Cycloadditions

Appropriate substitution of the two reaction partners may lead to a situation where each stereochemical mode of union can occur in two distinct regiochemical modes. As an example, we consider the $2\pi + 2\pi$ cyclodimerization of a monosubstituted olefin. For each stereochemical path, the preferred regiochemistry of the thermal reaction will be the one which minimizes the height of the ground surface barrier. The preferred regiochemistry of the photochemical reaction will be the one which minimizes the height of the lowest excited surface barrier and the energy gap involved in the decay process. The derivation of regioselection rules will be illustrated by reference to the thermal and photochemical nonionic $_2\pi_s + {}_2\pi_s$ and $_2\pi_s + {}_2\pi_a$ cycloadditions [26].

In the thermal $_2\pi_s + {}_2\pi_s$ case, the barrier height depends on three factors:

a) The Λ_3 intrapacket interactions which are of the $HO^D - HO^A$ and $LU^D - LU^A$ type.

b) The $\Lambda_1 - \Lambda_3$ interpacket interaction which involves a two electron matrix element. The regiochemical consequences of this interaction are neglected in our approach.

c) The $\Lambda_1 - \Lambda_2$ interpacket interactions which are of the $HO^D - LU^A$ and $HO^A - LU^D$ type. These interactions obtain whenever unsymmetrical olefins are involved and can become dominant.

The situation is considerably simplified in the case of thermal $_2\pi_s + _2\pi_a$ cycloaddition. Here, the barrier height is determined mainly by the $\Lambda_1 - \Lambda_2$ interpacket interactions which are of the $HO^D - LU^A$ and $HO^A - LU^D$ type.

Turning now our attention to photochemical reactions, regiochemical control can be exercised on the excited surface and on the decay process. In the photochemical $_2\pi_s + _2\pi_s$ case, the barrier height on the excited surface is determined by the following two factors:

a) The Λ_2 intrapacket interactions which are of the $HO^D - HO^A$ and $LU^D - LU^A$ type.

b) The $\Lambda_1 - \Lambda_2$ and $\Lambda_2 - \Lambda_3$ interpacket interactions which are of the $HO^D - LU^A$ and $HO^A - LU^D$ type. These interactions obtain whenever unsymmetrical olefins are involved and can become dominant.

On the other hand, the decay efficiency is determined exclusively by the $\Lambda_1 - \Lambda_3$ interpacket interaction, which is of the two electron type. Our approach neglects this factor and assumes that decay efficiency does not exert regiochemical control in nonionic $_2\pi_s + _2\pi_s$ cycloadditions.

In the case of $_2\pi_s + _2\pi_a$ photochemical cycloaddition, the barrier height on the excited surface depends on the following factors:

a) The $\Lambda_1 - \Lambda_2$ and $\Lambda_2 - \Lambda_3$ interpacket interactions which are of the $HO^D - LU^A$ and $HO^A - LU^D$ type.

b) The Λ_2 intrapacket interactions which are of the $HO^D - HO^A$ and $LU^D - LU^A$ type. These interactions obtain whenever unsymmetrical olefins are involved and can become dominant. On the other hand, the decay efficiency is determined by the $\Lambda_1 - \Lambda_2$ and $\Lambda_2 - \Lambda_3$ interpacket interactions which are of the $HO^D - LU^A$ and $HO^A - LU^D$ type.

We can now predict the reaction regioselectivity. For example, in $_2\pi_s + _2\pi_s$ thermal cycloadditions, the preferred regiochemistry will be ideally the one which maximizes the $HO^D - LU^A$, $HO^A - LU^D$, $LU^D - LU^A$ and $HO^D - HO^A$ interactions. The various regioselection rules, summarized in Table 5, warrant the following clarifying remarks:

a) The regiochemical condition specifies the MO interaction which should be maximized or minimized (minus sign in parenthesis) in order to optimize a barrier height or a decay process.

b) The regiochemical condition may involve maximization or minimization of an interaction matrix element which is of the $A\Gamma + B\Delta$ or $A\Gamma - B\Delta$ type. Accordingly, the types of interaction matrix elements involved are specified in Table 5.

c) $\Lambda_1 - \Lambda_2$ interpacket interactions are of the $HO^D - LU^A$ and $HO^A - LU^D$ type. The latter interaction is weaker and, thus, neglected. On the other hand, $\Lambda_2 - \Lambda_3$ inter-

Table 5. Regiochemical conditions for $2\pi + 2\pi$ cycloadditions[a]

Model reaction	Type of reaction	Barrier regiochemical condition		Decay regiochemical condition	
		$A\Gamma + B\Delta$	$A\Gamma - B\Delta$	$A\Gamma + B\Delta$	$A\Gamma - B\Delta$
$2\pi_s + 2\pi_s$	Δ, Nonionic	$HO^D - HO^A$, $\underline{LU^D - LU^A}$	$\underline{HO^D - LU^A}$	—	—
$2\pi_s + 2\pi_s$	Δ, Ionic	$HO^D - HO^A$, $\underline{LU^D - LU^A}$	$\underline{HO^D - LU^A}$	—	—
$2\pi_s + 2\pi_s$	$h\nu$, Nonionic[b]	$HO^D - HO^A$, $\underline{LU^D - LU^A}$	$HO^A - LU^D$, $\underline{HO^D - LU^A(-)}$	—	—
$2\pi_s + 2\pi_s$	$h\nu$, Ionic[b]	$HO^D - HO^A$, $\underline{LU^D - LU^A}$	$HO^A - LU^D$, $\underline{HO^D - LU^A(-)}$	—	$\underline{HO^D - LU^A(-)}$
$2\pi_s + 2\pi_a$	Δ, Nonionic	$\underline{HO^D - LU^A}$	—	—	—
$2\pi_s + 2\pi_a$	Δ, Ionic	$\underline{HO^D - LU^A}$	—	—	—
$2\pi_s + 2\pi_a$	$h\nu$, Nonionic[b]	$\underline{HO^D - LU^A(-)}$, $HO^A - LU^D$	$HO^D - HO^A$, $\underline{LU^D - LU^A}$	$\underline{HO^D - LU^A(-)}$, $HO^A - LU^D$	—
$2\pi_s + 2\pi_a$	$h\nu$, Ionic[b]	$\underline{HO^D - LU^A(-)}$, $HO^A - LU^D$	$HO^D - HO^A$, $\underline{LU^D - LU^A}$	$\underline{HO^D - LU^A(-)}$, $HO^A - LU^D$	—

[a] The symbol (−) denotes minimization of the corresponding overlap integral. Otherwise, maximization is implied. The key interactions, determined subject to the assumption stated in the text, are underlined.

[b] Assuming that D^*A^* lies above D^*A or DA^*.

packet interactions are of the $HO^A - LU^D$ type, assuming that the lowest diexcited diabatic surface is D^*A^*.

d) Λ_2 intrapacket interactions are of the $HO^D - HO^A$ and $LU^D - LU^A$ type. The former will become dominant if the DA^* diabatic surface lies lower in energy than the D^*A diabatic surface, and vice versa.

e) In all photochemical cases, a $HO^A - LU^D$ interaction will play a greater role than the $HO^D - LU^A(-)$ interaction if the D^+A^- diabatic surface is located closer to the D^*A^* diabatic surface. If the D^+A^- diabatic surface lies closer to the DA diabatic surface, the $HO^D - LU^A(-)$ interaction will exert greater influence.

f) In cases where the regiochemistry is determined conjointly by two or more interactions of different algebraic types (e.g., one of the $A\Gamma + B\Delta$ and another of the $A\Gamma - B\Delta$ type), the principal effect will be exerted by the interaction which is antipericyclic and couples the two diabatic surfaces closest in energy. When these conditions cannot be satisfied simultaneously, an estimate of the relative importance of the interactions becomes necessary.

On this basis, and assuming that intermediate formation is rate determining in $2\pi_s + 2\pi_s$ cycloadditions, the crucial interactions are those underlined in Table 5.

How are these regioselection rules expressed in terms of the HO and LU electron density properties of the two reactants [26]? As an illustration, consider the regiochemistry of the thermal $2\pi_s + 2\pi_a$ cyclodimerization of a monosubstituted ethylene, $CH_2=CHY$, which can occur in two distinct regiochemical ways, i.e., one involving HH union and the other HT union. The regioselection rule states that the preferred regiochemistry of a thermal $2\pi_s + 2\pi_a$ cycloaddition is the one which maximizes the $HO^D - LU^A$ interaction, i.e., the HO–LU interaction in the case of a cyclodimerization. We can write the HO and LU of the two reactants in terms of their constituent AO's.

$$HO = \ldots h_r x_r + h_s x_s \tag{4.24}$$

$$LU = \ldots l_t x_t + l_u x_u \tag{4.25}$$

Furthermore, we define:

$$\gamma_{rt} = \langle x_r | \hat{P} | x_t \rangle, \quad \gamma_{su} = \langle x_s | \hat{P} | x_u \rangle, \quad \text{etc.} \tag{4.26}$$

Accordingly, we may write:

$$\langle HO | \hat{P} | LU \rangle \, (HH) \propto (h_r l_t \gamma_{rt} + h_s l_u \gamma_{su}) \tag{4.27}$$

$$\langle HO | \hat{P} | LU \rangle \, (HT) \propto (h_r l_u \gamma_{ru} + h_s l_t \gamma_{st}) \tag{4.28}$$

The distance between the two olefins can be taken to be the same for HH and HT union. Since the uniting AO's are carbon 2p AO's, $\gamma_{rt} = \gamma_{su} = \gamma_{st} = \gamma_{ru} = \gamma$. We distinguish the following two cases:

a) $h_r \simeq h_t \simeq l_r \simeq l_t > h_s \simeq h_u \simeq l_s \simeq l_u$ or $h_r \simeq h_t \simeq l_r \simeq l_t < h_s \simeq h_u \simeq l_s \simeq l_u$.
A molecule which exhibits the relationship $h_r \simeq l_r > h_s \simeq l_s$, i.e., a molecule where the relative sizes of the AO coefficients of the two uniting atoms are similar in both HO and LU will be called a *homodensic* molecule. A typical example is that of butadiene. By substituting $h_r = h_t = l_r = l_t = a$ and $h_s = h_u = l_s = l_u = b$, the matrix elements become

$$\langle HO|\hat{P}|LU\rangle\,(HH) \propto (a^2 + b^2)\gamma \tag{4.29}$$

$$\langle HO|\hat{P}|LU\rangle\,(HT) \propto (2ab)\gamma \tag{4.30}$$

Since $a^2 + b^2 > 2ab$, we conclude that HH orientation will be preferred.

b) $h_r \simeq h_t \simeq l_s \simeq l_u < h_s \simeq h_u \simeq l_r \simeq l_t$ or $h_r \simeq h_t \simeq l_s \simeq l_u < h_s \simeq h_u \simeq l_r \simeq l_t$.
A molecule which exhibits the relationship $h_r \simeq l_s > h_s \simeq l_r$, i.e., a molecule where the relative sizes of the AO coefficients of the two uniting atoms are opposite in HO and LU, will be called a *heterodensic* molecule. A typical example is that of aminoethylene. By using the same substitution technique, we obtain

$$\langle HO|\hat{P}|LU\rangle\,(HH) \propto (2ab)\gamma \tag{4.31}$$

$$\langle HO|\hat{P}|LU\rangle\,(HT) \propto (a^2 + b^2)\gamma \tag{4.32}$$

We next consider the regiochemistry of thermal $_2\pi_s + {}_2\pi_s$ cyclodimerization. The crucial overlap integrals now take the form

$$HO\text{–}LU\,(HH) \propto a^2 - b^2 \tag{4.33}$$

$$HO\text{–}LU\,(HT) \propto 0 \tag{4.34}$$

Homodensic $CH_2 = CHY$

$$HO\text{–}HO \text{ or } LU\text{–}LU\,(HH) \propto a^2 + b^2 \tag{4.35}$$

$$HO\text{–}HO \text{ or } LU\text{–}LU\,(HT) \propto 2ab \tag{4.36}$$

$$HO\text{–}LU\,(HH) \propto 0 \tag{4.37}$$

$$HO\text{–}LU\,(HT) \propto a^2 - b^2 \tag{4.38}$$

Heterodensic $CH_2 = CHY$

$$HO\text{–}HO \text{ or } LU\text{–}LU\,(HH) \propto a^2 + b^2 \tag{4.39}$$

$$HO\text{–}HO \text{ or } LU\text{–}LU\,(HT) \propto 2ab \tag{4.40}$$

In the case of homodensic CH_2=CHY, HO–LU, HO–HO, and LU–LU matrix elements favor HH union. However, conflicting predictions arise in the case of heterodensic CH_2=CHY. Is there any way of deciding which of the two matrix elements will be more significant? The regiochemical preference for HT due to the HO–LU matrix elements is proportional to $a^2 - b^2$, while that for HH due to the HO–HO and LU–LU matrix elements is proportional to $(a - b)^2$. Since the ratio of the $(a - b)^2$ and $a^2 - b^2$ terms is smaller than unity, it follows that, *everything else being equal,* the HO–LU interactions will dominate the HO–HO and LU–LU interactions in determining the regioselectivity of the $_2\pi_s + _2\pi_s$ cyclodimerization of heterodensic CH$_2$=CHY.

We can easily generalize the analysis presented above. Thus, there are three types of interaction matrix elements, and each one enforces a different degree of regiochemical preference.

Algebraic Form of Matrix Element	Algebraic Form of Regiochemical Preference
AΓ + BΔ	$(a - b)^2$
AΓ	$a(a - b)$
AΓ − BΔ	$a^2 - b^2$

The AΓ − BΔ type interaction matrix element has the greatest regiochemical impact and the AΓ + BΔ the smallest.

While the algebraic manipulations outlined above are straightforward, a simpler recipe for determining regiochemical preferences can be offered. For example, in the case of the thermal $_2\pi_s + _2\pi_a$ cyclodimerization, the HO^D − LU^A interaction should be maximized. This can be accomplished by a geometric arrangement which joins the atom of one reaction partner having the highest HO electron density with the atom of the second reaction partner having the highest LU electron density. Obviously, before any specific predictions can be made, one should know the HO and LU electron densities of the reactants.

Table 6 conveys the results of computations pertaining to the relative sizes of the electron densities of the α and β carbons of substituted ethylenes of the type Y−C$_\alpha$H=C$_\beta$H$_2$. The relative sizes of the HO C$_\alpha$ and C$_\beta$ electron densities when Y is a pi acceptor depends on the type of calculation. MO electron densities of various molecular systems are given in Appendix I.

Table 7 summarizes the predictions of the regiochemistry of thermal $2\pi + 2\pi$ cyclodimerizations. Using the regioselection rules of Table 5 and data pertaining to the crucial AO coefficients of the HO's and LU's of the two cycloaddends, the regioselectivity of any $2\pi + 2\pi$ cycloaddition which proceeds in any given stereochemical manner can be predicted. Extension of these ideas to any cycloaddition is trivial [27, 28].

Table 6. Pi MO Electron densities (E.D.) of Y−C$_\alpha$H=C$_\beta$H$_2$

Molecular type	HO E.D.	LU E.D.	Y
Heterodensic A	$\alpha > \beta$	$\alpha < \beta$	NO$_2$
Heterodensic B	$\alpha < \beta$	$\alpha > \beta$	F, OH, NH$_2$
Homodensic A	$\alpha > \beta$	$\alpha > \beta$	−
Homodensic B	$\alpha < \beta$	$\alpha < \beta$	CN, Ph, CH=CH$_2$, CH$_3$

Table 7. Regioselectivity of thermal dimerization of $CH_2=CHY$

Reaction	Homodensic	Heterodensic	
2s + 2s, Δ	HH	HT	(HH?)
2s + 2a, Δ	HH	HT	
TB, Δ	HH	HT	

Unlike the case of HH vs. HT regioselectivity, which is a function of primary overlap, syn vs. anti regioselectivity depends upon secondary overlap which can contribute constructively or destructively to primary overlap. As an illustration, consider the regiochemistry of the thermal HH $_2\pi_s + _2\pi_s$ cyclodimerization of 1,3-butadiene. The HO–LU interaction matrix element will favor an anti geometry while the LU–LU (and HO–HO) interaction matrix element a syn geometry. These conclusions can be understood by reference to the drawings below [29].

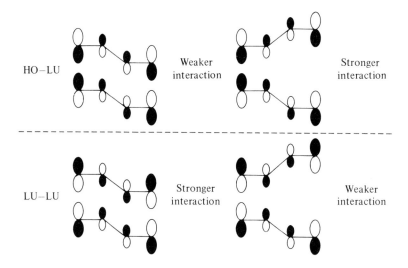

Once again, it should be pointed out that our discussions of regioselectivity were restricted to nonionic P2′ or P2′′ and ionic P2′′ $2\pi + 2\pi$ photocycloadditions. Extension of the analysis to other types of photocycloadditions is trivial.

4.6 Isomeric Reactions and the Topochemistry of $2\pi + 2\pi$ Cycloreversions

The concept of reaction polarity is useful for making predictions regarding rate variations within a graded series of chemical reactions. Thus, for example, we can say that, in the reaction series shown below, relative reactivity will vary in the order *1 > 2 > 3* reflecting a similar variation in reaction polarity. By contrast, polarity does not provide

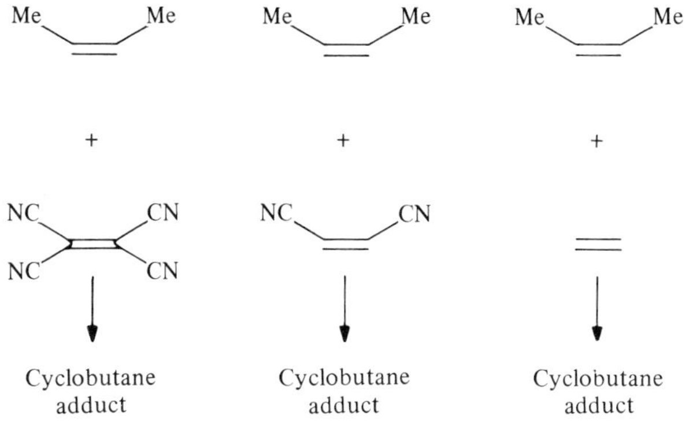

Me   Me   Me   Me   Me   Me

+   +   +

NC   CN   NC   CN

NC   CN

Cyclobutane   Cyclobutane   Cyclobutane
adduct   adduct   adduct

1   *2*   *3*

the basis for predicting the relative rate of two isomeric multicentric reactions, e.g., the reaction of **A** plus **A** and the reaction **B** plus **C** shown below. This can be inferred from the fact that there are two constraints placed on these two reactions, i.e., one at the product level, where the cycloadduct is common to both reactions, and one at the neighborhood of the transition state, where the pi type state manifolds for the complexes are nearly identical[1]. Accordingly, the relative rates of the two isomeric reac-

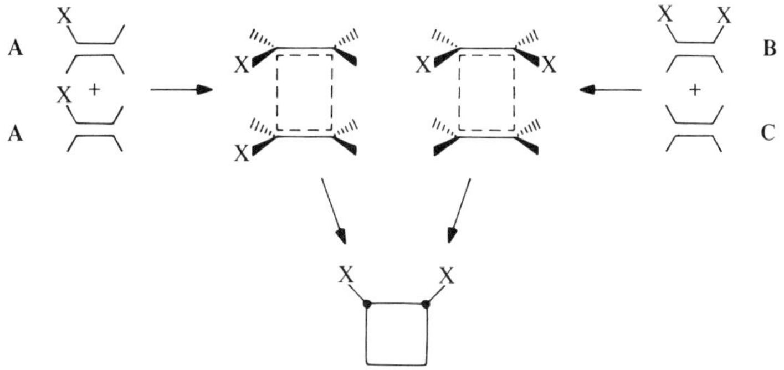

tions will depend on the relative stabilities of the ground state reactants. There ideas are illustrated diagrammatically in Fig. 31. Clearly, in isomeric multicentric reactions of the type shown above **A** plus **A** will react faster or slower than **B** plus **C** depending on their relative thermodynamic stability. On the other hand, sigma bond strengths will dictate the preferred fragmentation of the cycloadduct. By reference to [21] we predict that decomposition to **A** + **A** will be favored, at least in most cases.

---

1 An example will serve to illustrate how this electronic constraint arises. Thus, the pi MO manifold of square 1,2-dicyanocyclobutadiene should be the same whether it is constructed from the pi MO's of the HC=CH and NCC=CCN fragments or the pi MO's of two NCC=CH fragments.

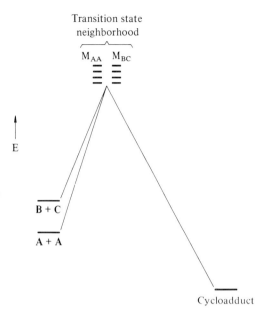

Fig. 31. Simplified energy diagram for thermal isomeric reactions **A + A** and **B + C**. The pi state manifolds M_{AA} (**A + A** reaction) and M_{BC} (**B + C** reactions) become nearly identical in the neighborhood of the transition state

It should be emphasized that the above discussion is pertinent for isomeric reactions which occur in a $_2\pi_s + {}_2\pi_s$ manner. Furthermore, the electronic constraints in the neighborhood of the transition state dictate that two isomeric reactions may be of the same or different ionicity type, e.g., both ionic, or, one pseudoionic and the other ionic, etc.

4.7 The Topochemistry of Intramolecular Cycloadditions

A molecule which contains more than one unsaturated fragment constitutes a convenient system for studying the toposelectivity of a cycloaddition constrained to proceed via a well-defined stereochemical path. For example, in the model system shown below, there is a competition among three well-defined $_2\pi_s + {}_2\pi_s$ paths.

The following predictions can be made:

1. The preferred thermal path will be the one which involves the best donor-acceptor match, if the triplet excitation energies of the **A**, **B**, and **C** fragments are comparable.

2. In intermolecular photocycloadditions, the two reactants have to approach from infinity to the final bonding distance. In reactions of high polarity, an excited intermediate is formed at some intermediate distance and significant energy wastage may render the lowest energy reactive path unproductive. By contrast, in most intramolecular photocycloadditions, the two reacting fragments are constrained to be in the vicinity of each other and excited intermediate formation is avoided. In addition, the skeletal restrictions prohibit the introduction of a sufficient number of donor and/or

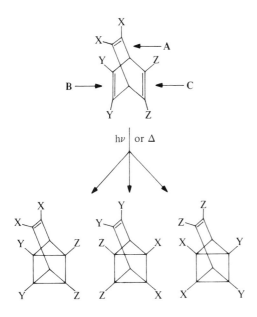

acceptor groups to render the reaction highly polar. In short, *dual channel mechanisms (vide infra)* are probably inoperative in intramolecular photocycloadditions and the toposelectivity of an intramolecular photocycloaddition can be predicted by identifying the lowest energy excited state and its bonding characteristics. If the $\pi\pi^*$ singlet excitation energies of **A**, **B**, and **C** fragments are comparable, the preferred path will be the one which involves the best donor-acceptor match, assuming that interaction matrix elements do not differ substantially[1].

4.8 The Selectivity-Polarity Relationship

How does the selectivity of a thermal $2\pi + 2\pi$ cycloaddition depend on reactivity, or, in other words, on polarity? Because of the entirely different shapes of the P.E. surfaces for nonionic and ionic cycloadditions, the selectivity-polarity relationship for each of the two reaction classes will be discussed separately.

a) Stereoselectivity-polarity relationship in thermal nonionic cycloadditions: In this case, one may inquire as to how the ratio of the rate of the $_2\pi_s + _2\pi_a$ cycloaddition over the rate of the $_2\pi_s + _2\pi_s$ cycloaddition depends on polarity. However, since the

---

1 These predictions were implicit in the static LCFC treatment of photocycloadditions outlined in [6e] (in particular, see Fig. 4). For a misinterpretation of this work, see: Bender, C. O., Wilson, J., Helvetica Chimica Acta *59*, 1469 (1976).

transition states of the two paths are determined by different types of packet interactions, no simple analysis is possible.

b) Regioselectivity-polarity relationship in thermal nonionic cycloadditions. In $_2\pi_s$
$+ _2\pi_a$ cycloadditions, the rate ratio for two different regiochemical unions will increase
as polarity increases and the DA and D^+A^- diabatic surfaces interact more strongly.

We consider this proposition in some detail. We define the following terms:

$$P_A = \langle D^+A^- | \hat{P} | DA \rangle_A \quad \text{(regiochemistry A)} \tag{4.41}$$

$$P_B = \langle D^+A^- | \hat{P} | DA \rangle_B \quad \text{(regiochemistry B)} \tag{4.42}$$

Furthermore, we assume that $|P_A| > |P_B|$, i.e., A is the preferred regiochemical path.
The height of the barrier for regiochemistry A is a function of the quantity $P_A^2/(I_D - A_A)$,
which is a measure of the strength of the interaction of the DA and D^+A^- diabatic surfaces. Similarly, the height of the barrier for regiochemistry B is a function of the quantity $P_B^2/(I_D - A_A)$. Regioselectivity will be a function of the difference of the two quantities and will increase as polarity increases, assuming that P_A and P_B remain constant.

In the case of $_2\pi_s + _2\pi_s$ cycloaddition, the problem becomes complicated since
regioselectivity depends upon Λ_3 intrapacket as well as $\Lambda_1 - \Lambda_2$ interpacket interactions and an increase in polarity can affect both types of interactions.

c) Stereoselectivity-polarity relationship in ionic cycloadditions. In this case, an approximation of the dependence of the ratio of the rate of the $_2\pi_s + _2\pi_a$ cycloaddition over
the rate of the $_2\pi_s + _2\pi_s$ cycloaddition on reaction polarity can be made by assuming
that the barrier height of each reaction type depends solely on the interaction of the
DA and D^+A^- crossing diabatic surfaces, which, in turn, depends solely on the
$\langle DA | \hat{P} | D^+A^- \rangle$ matrix element.

We define the following terms:

$$P_A = \langle D^+A^- | \hat{P} | DA \rangle_A \quad (_2\pi_s + _2\pi_a) \tag{4.43}$$

$$P_B = \langle D^+A^- | \hat{P} | DA \rangle_b \quad (_2\pi_s + _2\pi_s) \tag{4.44}$$

Stereoselectivity will be a function of the difference of these two quantities, S.

$$S = |P_A| - |P_B| \tag{4.45}$$

If we assume that $P_B \simeq 0$, the above equation becomes $S = P_A$. Now, P_A is a product
of coefficient terms and AO overlap integral terms. As polarity increases, and the intersection of the DA and D^+A^- occurs at progressively looser geometries, S decreases
because spatial AO overlap also decreases. *Increasing polarity in ionic cycloadditions*
will narrow the energy gap separating the $_2\pi_s + _2\pi_s$ and $_2\pi_s + _2\pi_a$ stereochemical paths.
The above result has important chemical consequences. *It suggests that, in ionic cyclo-*
additions, the energy separation of LM and NLM paths imposed by electronic effects
is small and, hence, may become dominated by other effects.

d) Regioselectivity-polarity relationship in ionic cycloadditions. In ionic $_2\pi_s + _2\pi_a$ cycloadditions, the regioselectivity of the reaction depends solely on the difference of two $\langle D^+A^- | \hat{P} | DA \rangle$ matrix elements corresponding to two different regiochemical union modes. This difference is a function of spatial overlap, Accordingly, *as reaction polarity increases and the DA $-$ D$^+$A$^-$ intersection occurs at progressively looser geometries, the regioselectivity of ionic $_2\pi_s + _2\pi_a$ cycloadditions will decrease.*

In ionic $_2\pi_s + _2\pi_s$ cycloadditions, the regioselectivity of the reaction may be assumed to depend solely on the difference of two $\langle D^+A^- | \hat{P} | DA \rangle$ matrix elements corresponding to two different regiochemical modes of union. The conclusions are the same as in the $_2\pi_s + _2\pi_a$ case.

4.9 Reaction Intermediates and Types of Organic Mechanisms

The intermediates which can be encountered in $2\pi + 2\pi$ cycloadditions are listed in Table 8. The commonly used terminology is indicated.

In all previous discussions, we have made the tacit assumption that the rate determining step in thermal ionic and photochemical nonionic cycloadditions, where efficient decay to ground products is possible, is the one leading to intermediate formation. Obviously, other possibilities exist and a general classification of reaction mechanisms becomes necessary. Such a classification can be implemented by reference to Fig. 32 which shows four possible reaction profiles that can be deduced on the basis of our theoretical approach. We distinguish the following different mechanistic types:

a) Type A reactions. In this case, the reaction rate depends on E_1. Thermal ionic reactions and P1'' and P2' nonionic photoreactions can be typical examples.

b) Type A' reactions. As in (a), the rate depends on E_1. Thermal nonionic reactions and most Pl' nonionic photoreactions are typical examples.

c) Type B reactions. In this case, the rate depends on the stability of I and also on E_2. Thermal ionic reactions and P1'' and P2' nonionic photoreactions can be typical examples.

d) Type C reactions. As in (c), the rate depends on the stability of I and also on E_2. Thermal superionic reactions and P2'' nonionic photoreactions are typical examples.

The reader will appreciate that, while our approach allows for a differentiation among type A', type A *or* B, and type C reactions, it cannot differentiate in an *a priori* sense between type A and type B reactions. In general, the regiochemical conditions for minimization of E_1 and E_2 and maximization of intermediate stability are the same for all mechanistic types in most, but not all, reactions. Exceptions arise in thermal ionic antiaromatic and photochemical nonionic aromatic reactions where depending upon mechanistic type, A or B, a different regiochemical bias may be exerted. For example, in a thermal ionic antiaromatic reaction, E_1 and intermediate stability are

Table 8. Reaction intermediates in $2\pi + 2\pi$ cycloadditions

Reaction type	Intermediate designation	Literature nomenclature
$2\pi_s + 2\pi_s$, Δ, Nonionic	**D** . . . **A** or N	Contact charge transfer complex Ground charge transfer complex
$2\pi_s + 2\pi_a$, Δ, Nonionic	**D** . . . **A** or N	
$2\pi_s + 2\pi_s$, Δ, Ionic	**D** . . . **A** or N ___ M or N^*	 "Dipolar intermediate"
$2\pi_s + 2\pi_a$, Δ, Ionic	**D** . . . **A** or N ___ N^*	
$2\pi_s + 2\pi_s$, hν, Nonionic (Excited surface only)	**D** . . . **A*** (**D*** . . . **A**) or M^* ___ M ___ Ξ'^*	Encounter complex Exciplex No agreed-upon term
$2\pi_s + 2\pi_a$, hν, Nonionic (Excited surface only)	**D** . . . **A*** (**D*** . . . **A**) or M^* ___ O or N^*	 No existing term Excited charge transfer complex
$2\pi_s + 2\pi_s$, hν, Ionic (Excited surface only)	**D** . . . **A*** (**D*** . . . **A**) or M^* ___ N'^*	 No existing term
$2\pi_s + 2\pi_a$, hν, Ionic (Excited surface only)	**D** . . . **A*** (**D*** . . . **A**) or M^* ___ N'^*	

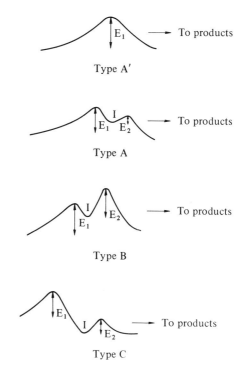

Type A'

Type A

Type B

Type C

Fig. 32. General classification of reaction
mechanisms

subject to $HO^D - LU^A$ control, while E_2 is subject to $HO^A - LU^D$ control. All three reactivity parameters are subject to indirect $HO^D - HO^A$ (or, $LU^D - LU^A$) control. Thus, a type A mechanism is compatible with $HO^D - LU^A$ control and a type B mechanism with $HO^D - LU^A$ or $HO^A - LU^D$ control.

The effect of polarity on E_1 and intermediate stability is independent of mechanistic type, assuming that the polarity change itself, rather than the change in the interaction of the diabatic surfaces, determines primarily the energetic readjustment of the adiabatic surfaces. A problem arises in reactions of the type mentioned above where an increase in polarity leads to a reduction of E_1 and an increase of E_2 and the intermediate stability. Thus, a type A mechanism is compatible with a rate increase accompanying an increase in polarity, while a type B mechanism is compatible with either an increase or a decrease. Table 9 amounts to a comparative study of type A, A', and B thermal reactions.

On the basis of the above considerations and for the purpose of analyzing the experimental data, we shall assume as a working hypothesis that all thermal nonionic reactions belong to type A', all thermal ionic reactions to type A, all thermal superionic reactions to type C, and all photochemical nonionic reactions to type A or A'. However, the reader should not lose sight of the main objective of this work which is to develop and illustrate a general procedure for solving problems of chemical reactivity. Indeed, the stated assumptions are hardly expected to be valid for all chemical reactions and one should always be alert to identify cases where a breakdown occurs.

Table 9. Reactivity trends of $_2\pi_s + _2\pi_s$ and $_2\pi_s + _2\pi_a$ cycloadditions[a]

Model reaction	Type of reaction	Stereoselectivity response	Regioselectivity response
$_2\pi_s + _2\pi_s$	Δ, Nonionic type A'	−	−
$_2\pi_s + _2\pi_s$	Δ, Ionic type A	↑	↓
$_2\pi_s + _2\pi_s$	Δ, Ionic type B	−	↓
$_2\pi_s + _2\pi_a$	Δ, Nonionic type A'	−	↑
$_2\pi_s + _2\pi_a$	Δ, Ionic type A	↓	↓
$_2\pi_s + _2\pi_a$	Δ, Ionic type B	−	↓

[a] A downwards pointing arrow indicates a decrease and an upwards pointing arrow an increase as polarity increases.

4.10 $2\pi + 2\pi$ Nonionic Photocycloadditions Involving $n\pi^*$ Excitation

The elucidation of the mechanism of $n\pi^*$ photocycloadditions poses a great challenge to the theorist. In such reactions, the cycloaddition complex begins its travel on the adiabatic surfaces with a singly occupied n orbital and ends to the ground state product resulting from pi fusion of the two reactants and a regeneration of a doubly occupied n orbital. Accordingly, the efficiency of an $n\pi^*$ photocycloaddition will depend upon the extent of fulfillment of the following conditions:

a) Optimization of pi intrapacket interactions. This guarantees efficient decay to ground cycloadduct via the "hole" created by the $\Lambda_1 - \Lambda_3$ two electron interaction.

b) Optimization of the photochemical barrier, i.e., optimization of the interaction of the lowest energy charge transfer and $n\pi^*$ locally excited diabatic surfaces.

c) Optimization of an inter- or intrapacket interaction which can become responsible for electron transfer to the singly occupied n orbital.

The construction of adiabatic P.E. surfaces for photocycloadditions involving $n\pi^*$ is straightforward but laborious. In this case, we need to construct the Λ_1, Λ_2 and Λ_3 packets of diabatic surfaces by permuting six electrons among five MO's.

Thus, we have chosen to convey the essential aspects of $n\pi^*$ photocycloadditions by employing an alternative symbolic convention. The procedure involves the following steps:

a) Specification of the various types of inter- and intrapacket interactions.

b) Identification of the diabatic surfaces which lead via crossing from reactants to products.

c) Prediction of choroselectivity. Specifically, the preferred chorochemistry will optimize simultaneously pi intrapacket interactions, the interaction of the lowest energy $n\pi^*$ locally excited and charge transfer diabatic surfaces connected with the photochemical barrier, and, finally, the diabatic surface interaction which allows back electron transfer to the singly occupied n orbital.

Consider the nonpolar reactions shown below, where it is assumed that the olefin partner acts as the pi acceptor and the carbonyl as the n donor.

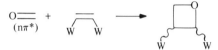

The inter- and intrapacket interactions are spelled out below.

Intrapacket interactions		Interpacket interactions
–	← Λ_1 ←	
		HO–LU
		LU–HO
		n–LU
HO–HO		
LU–LU	← Λ_2 ←	
n–HO		
		HO–LU
		LU–HO
		n–LU
HO–HO		
LU–LU	← Λ_3 ←	
n–HO		

The crucial crossing surfaces are D*A, D^+A^- and the principal component of the Λ_3 boundary which interacts in a one electron sense with D^+A^-. The reaction progress is exemplified by the diagram shown in Fig. 33.

At the first crossing, the reaction system surmounts the photochemical barrier, while at the second crossing it makes a transition to Λ_3 leading to product. The crucial orbital interactions which must be simultaneously optimized are $HO^D - HO^A$, $LU^D - LU^A$ and $n^D - HO^A$. The predicted geometry of approach is shown below, i.e., a sideways approach is favorable and a coaxial approach is unfavorable. The reaction will be $_2\pi_s + _2\pi_s$ stereoselective.

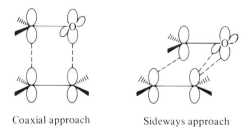

Coaxial approach   Sideways approach

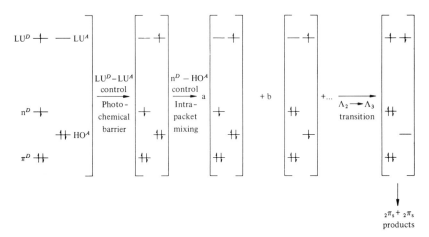

Fig. 33. Schematic representation of the mechanism of nonpolar $2\pi + 2\pi$ photocycloadditions initiated by $n\pi^*$ excitation

The same conclusions will be reached even if DA^* is not crossed by D^+A^- as long as D^+A^- is lower energetically than D^-A^+. In this case, the mere mixing of diabatic surfaces will enforce the same optimization conditions for the photochemical barrier as well as the transition to Λ_3 leading to products. Finally, the treatment can be extended to the reaction shown below. By following the same line of reasoning, we predict both oxetane and cyclobutane formation to be $_2\pi_s + _2\pi_s$ stereoselective and oxetane formation to be more efficient if regeneration of the doubly occupied n orbital is a key feature of the reaction.

Next, we consider the polar reaction shown below, where it is assumed that the olefin partner acts as the pi donor and the carbonyl as the pi acceptor.

The crucial crossing surfaces are DA^*, D^+A^- and the principal component of the Λ_3 boundary which interacts in a one electron sense with D^+A^-. The reaction progress is exemplified by the diagram shown in Fig. 34. The crucial orbital interactions which must be simultaneously optimized are HO–HO, LU–LU and n–HO. Once again, we predict a favorable sideways approach and an unfavorable coaxial approach. The reaction will be $_2\pi_s + _2\pi_s$ stereoselective.

90

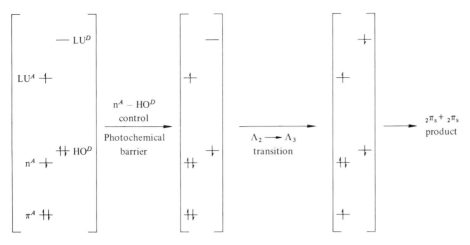

Fig. 34. Schematic representation of the mechanism of polar $2\pi + 2\pi$ photocycloadditions initiated by $n\pi^*$ excitation

The reader should verify that the same conclusions will be reached even if DA* is not crossed by D^+A^-, as long as D^+A^- is energetically lower than D^-A^+. Once again, the treatment can be extended to the reaction shown below. Both oxetane and cyclobutane products are expected to be formed in a $_2\pi_s + _2\pi_s$ manner with oxetane the preferred product under the assumptions stated before.

The stereoselectivity of $n\pi^*$ photocycloadditions will be $_2\pi_s + _2\pi_s$ irrespective or reaction polarity. However, on the assumption that the reaction regiochemistry is barrier controlled, regioselectivity will be consistent with maximal $LU^D - LU^A$ interaction in nonpolar cases and maximal $n^A - HO^D$ interaction in polar cases.

Finally, an important proviso should be added. Specifically, in polar $n\pi^*$ photocycloaddition decay to ground state from a photononaromatic or photoantiaromatic geometry may not be prohibitively unfavorable. In such cases, the reaction stereochemistry will be consistent not with a simultaneous optimization of n–HO, HO–HO, and LU–LU interactions but rather with maximization of the n–HO interaction controlling the height of the photochemical barrier and the reaction will proceed via a diradicaloid intermediate and will be nonstereoselective.

The classification of photocycloadditions initiated by $n\pi^*$ excitation can be effected by applying the inequalities developed in the case of photoreactions initiated by $\pi\pi^*$ excitation, with $G(\pi\pi^*)$ replaced by $G(n\pi^*)$. In addition, the analysis of the interaction of an $n\pi^*$ excited carbonyl with a pi bond can be extended to the interaction of an $n\pi^*$ excited carbonyl with a sigma bond.

4.11 The Effect of Low Lying Diexcited Diabatic Surfaces in Photoreactions

A diexcited configuration of the type D^*A^* may attain very low energy; polyenes having low lying diexcited states are known [30]. Accordingly, further attention should be paid to the role that diexcited diabatic surfaces play in determining photoreactivity, especially in the more ubiquitous nonionic P2$'$ or P2$''$ $2\pi + 2\pi$ cycloadditions.

Let us briefly review the key features of nonionic photochemical $2\pi + 2\pi$ cycloadditions. In $_2\pi_s + _2\pi_s$ cycloadditions, the height of the photochemical barrier and the stability of the excited intermediate M depend mainly upon the primary interaction of DA^* (or D^*A) and D^+A^-, which is of the $HO^D - HO^A$ (or $LU^D - LU^A$) type. If D^*A^* approaches DA^* (or D^*A) in energy, the interaction of this diabatic surface with the D^+A^- diabatic surface will have to be considered. This interaction is of the $HO^A - LU^D$ type and will be zero for symmetrical systems but may become appreciable for unsymmetrical systems. In the latter case, the M intermediate will have a strong diexcitation contribution.

In photochemical nonionic $_2\pi_s + _2\pi_a$ cycloadditions, the height of the photochemical barrier and the stability of the excited intermediate depend upon the interaction of the DA and D^+A^- diabatic surfaces (symmetrical or unsymmetrical cases). The additional $D^+A^- - DA^*$ and $D^+A^- - D^*A$ interactions come to play in unsymmetrical systems. However, if D^*A^* lies low in energy, the $D^*A^* - D^+A^-$ interaction can dominate the $DA - D^+A^-$ interaction.

The above considerations are pictorially illustrated in Fig. 35, where the relative energies of the pertinent configurations have been evaluated at moderate intermolecular distances. The specific predictions are collected in Table 10, where a primary interaction is defined as the one between D^+A^- and the lowest excited configuration while a secondary interaction is defined as the one between D^+A^- and the next higher excited configuration.

An interesting situation arises when a diexcited configuration of the D^*A^* or DA^{**} type attains lower energy than the lowest energy monoexcited DA^* configuration at infinite intermolecular distance. In such cases, the efficiency of decay to the lowest adiabatic excited surface will have to be considered as well. For example, when D^*A^* attains a lower energy than DA^* and D^*A the excited diabatic surfaces become as shown in Fig. 36. Here, the E^* barrier is optimized by maximizing the $HO^D - HO^A$ interaction, the decay efficiency at point A by minimizing the $HO^A - LU^D$ interaction, and the energy of the lowest adiabatic surface past point A by maximizing the $HO^A - LU^D$ interaction. These considerations can be illustrated pictorially by means of Fig. 35. The specific predictions are collected in Table 10. Other possible situations can be analyzed similarly.

A close scrutiny of Fig. 35, leads to the following generalizations:

a) As long as the monoexcited configurations of the D^*A and DA^* type lie underneath the diexcited configuration of the D^*A^* type at infinite intermolecular distance, e.g., cases I, II, and V, the various conclusions regarding the principal features of the lowest excited adiabatic surface discussed before will remain valid. By contrast, cases III, IV, and VI are not simple to deal with because the decay condition conflicts with the con-

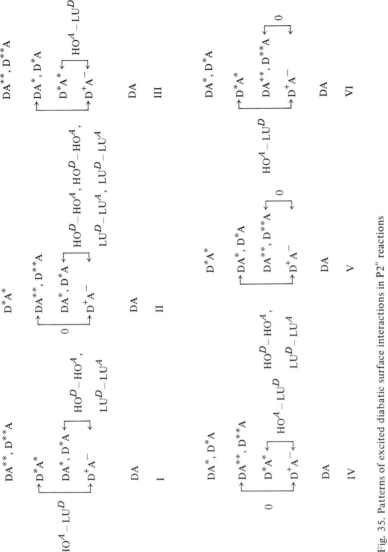

Fig. 35. Patterns of excited diabatic surface interactions in P2″ reactions

dition for optimization of the lowest excited surface. Fortunately, these three cases are not common.

b) The complications encountered in singlet photoreactions are drastically diminished in triplet photoreactions. In such cases, diexcited diabatic surfaces lie well above monoexcited diabatic surfaces. Hence, the intimacies of the lowest triplet excited surface can be revealed by simple considering the $^3DA^*$, $^3D^*A$, $^3D^+A^-$, and $^3D^-A^+$ diabatic surfaces.

Table 10. Bonding and regiochemical conditions for various patterns of excited diabatic surface interactions[a]

Photoreaction type[b]	Primary pericyclicity	Secondary pericyclicity	Primary[c] regiochemical control	Secondary[c] regiochemical control	Decay[d] regiochemical control
I, $2\pi_s + 2\pi_s$	P	NP	$\begin{cases} HO^D - HO^A, \\ LU^D - LU^A \end{cases}$	$HO^A - LU^D$	–
I, $2\pi_s + 2\pi_a$	NP	P			
II, $2\pi_s + 2\pi_s$	P	NP	$\begin{cases} HO^D - HO^A, \\ LU^D - LU^A \end{cases}$	None	–
II, $2\pi_s + 2\pi_a$	NP	NP			
III, $2\pi_s + 2\pi_s$	NP	P	$HO^A - LU^D$	$\begin{cases} HO^D - HO^A, \\ LU^D - LU^A \end{cases}$	$HO^A - LU^D(-)$
III, $2\pi_s + 2\pi_a$	P	NP			
IV, $2\pi_s + 2\pi_s$	NP	NP	$HO^A - LU^D$	None	$HO^A - LU^D(-)$
IV, $2\pi_s + 2\pi_a$	P	NP			
V, $2\pi_s + 2\pi_s$	NP	P	None	$\begin{cases} HO^D - HO^A, \\ LU^D - LU^A \end{cases}$	None
V, $2\pi_s + 2\pi_a$	NP	NP			
VI, $2\pi_s + 2\pi_s$	NP	NP	None	$HO^A - LU^D$	$HO^A - LU^D(-)$
VI, $2\pi_s + 2\pi_a$	NP	P			

a The symbol (–) denotes minimization of the corresponding overlap integral. Otherwise, maximization is implied.   P = pericyclic, NP = nonpericyclic.

b See Fig. 35 for definitions of symbols I, II, . . . VI.

c With respect to initial photochemical barrier.

d With respect to decay to the lowest excited adiabatic surface.

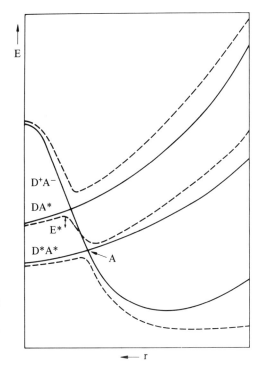

Fig. 36. Excited P.E. surfaces for the $2\pi_s + 2\pi_a$ cycloaddition of D and A^* where the lowest energy excited diabatic surface is D^*A^*. Solid lines indicate diabatic and dashed lines adiabatic surfaces. It is assumed that the two reactants are unsymmetrical

4.12 The Problem of Energy Wastage. The Concept of Dual Channel Photoreactions

In a normal photochemical reaction, the two reacting molecules or fragments descend towards or remain on the lowest energy excited surface on their way towards excited product or eventual decay to ground product. *Dual channel* photochemical reactions are also possible. The key features of such reactions are the following:

a) The lowest energy path is responsible for inefficient product formation and efficient energy wastage due to decay of an excited intermediate back to ground state reactants.

b) A higher energy path is responsible for efficient product formation and little energy wastage.
Such dual channel mechanisms are expected in highly polar nonionic reactions.

The diabatic surfaces of the model $2\pi + 2\pi$ cycloaddition reactions suggest that whenever the minimum of the D^+A^- approaches DA (i.e., polar nonionic cycloadditions), decay of the corresponding excited intermediate to ground reactants will become responsible for energy wastage and reduced reaction efficiency. The diabatic surface interactions which minimize the energy of this intermediate should dictate energy wastage and inefficient product formation, while those which produce a higher energy excited intermediate should dictate reduced energy wastage and efficient product formation. *π + π cycloadditions are ideal for testing this hypothesis since different stereochemical and regiochemical modes of union signal different diabatic surface interactions.*

There are two important mechanisms by which a ground state and an excited reactant, which constitute a good donor-acceptor pair by reference to ground state properties, may avoid return to the ground surface and proceed to yield products. These are discussed in detail below.

I. nπ* State – Dual Channel Mechanism in 2π + 2π Photocycloadditions

A molecule **D** reacting with a molecule **A***, which is excited in an $n\pi^*$ sense, has the following options:

a) Reaction via a geometry which maximizes the $DA^* - D^+A^-$ and minimizes the $DA - D^+A^-$ interaction. This minimizes the energy of the excited intermediate and gives rise to energy wastage and inefficient product formation.

(a)

(b)

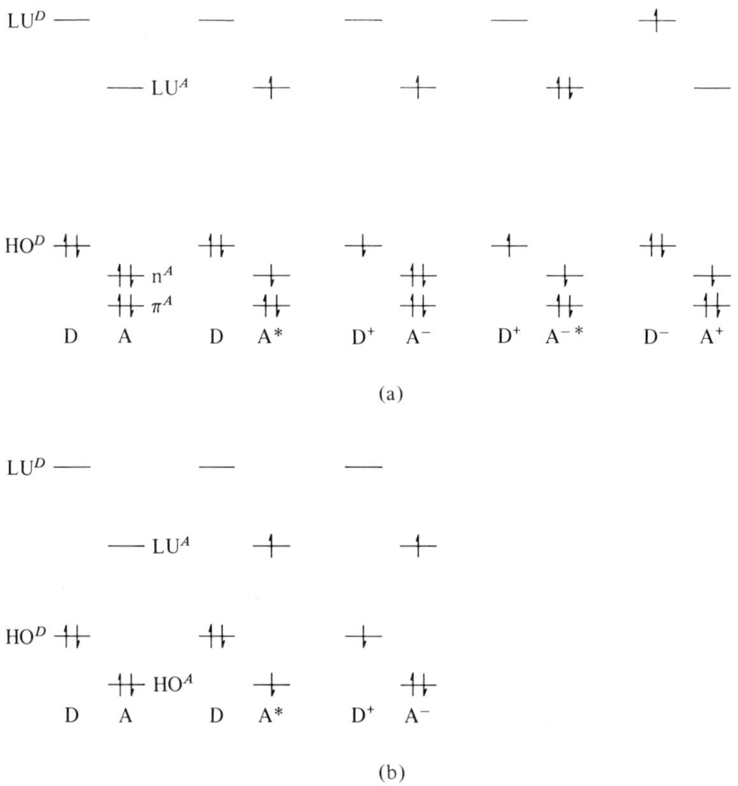

Fig. 37. Basis set configurations for dual channel photocycloadditions initiated by a) $n\pi^*$ and b) $\pi\pi^*$ photoexcitation

b) Reaction via a geometry which maximizes the $DA - D^+A^-$ and minimizes the $DA^* - D^+A^-$ interaction. This maximizes the energy of the excited intermediate, reduces energy wastage, and allows the reaction to proceed efficiently via a higher energy path.

The geometry which maximizes the $DA - D^+A^-$ interaction also maximizes the $DA^* - D^+A_n^-{}^*$ interaction, where $D^+A_n^-{}^*$ is a charge excited diabatic surface which can mix with the locally excited diabatic surface and has lower energy than the $D^-A_n^+$ charge transfer diabatic surface. The various configurations are shown in Fig. 37 and the concepts are illustrated by means of Fig. 38. Typical reactions where this dual mechanism may arise are the photocycloadditions of $n\pi^*$ excited carbonyls and donor olefins.

II. $\pi\pi^*$ State – Dual Channel Mechanism in $2\pi + 2\pi$ Photocycloadditions

Consider a molecule **D** reacting with a molecule **A*** which is excited in a $\pi\pi^*$ sense. The reaction system has the following options:

a) Reaction via a geometry which maximizes the $DA^* - D^+A^-$ interaction and minimizes the $DA - D^+A^-$ interaction. This leads to energy wastage and inefficient product formation.

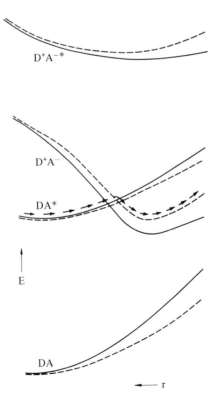

Fig. 38. Higher energy $2\pi_s + 2\pi_a$ path in dual channel $2\pi + 2\pi$ photocycloaddition initiated by $n\pi^*$ excitation. Solid lines indicate diabatic and dashed lines adiabatic surfaces

b) Reaction via a geometry which minimizes the $DA^* - D^+A^-$ interaction and maximizes the $DA - D^+A^-$ interaction. In this case, the energy separation of the excited intermediate and the ground adiabatic surface is greater, making energy wastage less probable and leading to efficient product formation. The various configurations are shown in Fig. 37 and the concepts are illustrated by means of Fig. 39.

The results of the analysis presented above may be summarized by the following statement: polar photocycloadditions involving dual channel mechanisms will proceed in a chorochemical manner expected for the corresponding thermal reactions on the basis of the Woodward-Hoffmann rules, i.e., the chorochemistry of those reactions will be "anti-Woodward-Hoffmann".

Two other interesting implications of energy wastage are the following:

a) In Sect. 4.2, we discussed the effect of polarity on the rate of photocycloadditions *assuming a minimal rate of decay of the excited intermediate to ground state reactants.* In this section, we have argued that in any highly polar photoreaction this process can become prevalent. These ideas can now be synthesized into a general predictive rule: *the rate of a photoreaction will pass through a maximum and subsequently decline to a minimum as polarity increases, for fixed stereochemistry.*

b) A singlet reaction will be more prone towards energy wastage than a triplet reac-

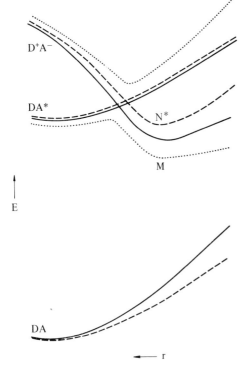

Fig. 39. Dual channel mechanism of $2\pi + 2\pi$ photocycloaddition initiated by $\pi\pi^*$ excitation. Solid lines indicate diabatic surfaces. Dashed lines indicate adiabatic surfaces appropriate to $_2\pi_s + _2\pi_a$ photoreactions. Dot lines indicate adiabatic surfaces appropriate to $_2\pi_s + _2\pi_s$ photoreaction. Energy wastage is due primarily to decay of M

tion because in the former case there is no multiplicity prohibition for decay of the singlet excited intermediate to ground reactants.

Finally, certain cautionary remarks are in order:

a) Most photoreactions which proceed via a dual channel mechanism are expected to be *overall* inefficient due to incursion of the path leading to energy wastage.

b) Whenever maximization of one interaction and minimization of a second interaction of diabatic surfaces cannot be simultaneously achieved, the crucial interaction becomes the one which couples the two diabatic surfaces which are separated by the smallest energy gap.

c) In formulating the concept of dual channel mechanism, we have assumed that the major reaction product is the one which is consistent with maximizing the energy gap between an excited surface on which reaction is initiated and the ground surface so that decay of the excited intermediate leading to energy wastage is minimized. This formulation ignores the effect of decay efficiency to ground state products past the energy wasting intermediate. However, in polar nonionic reactions, decay from the lowest excited to the ground surface is expected to be an intrinsically favorable process along the reaction coordinate due to a relatively small energy gap.

d) Although, for the purpose of illustration, we have assumed that the minimum of D^+A^- lies below that of DA^* or D^*A, this is not a necessary condition. Thus, a dual channel mechanism can be operative even if the reverse situation obtains and DA^* or D^*A has lower energy. The important thing to remember is that such a mechanism depends on the energetic proximity of the ground and lowest excited electronic states at long intermolecular distance. An increase in polarity leads to an increase in the bias towards a dual channel mechanism by contributing directly or indirectly to minimization of the corresponding energy gap.

4.13 Generalizations

The following general statements can now be made:

a) A thermal $_2\pi_s + _2\pi_s$ cycloaddition constitutes a model for a thermal antiaromatic reaction while a photochemical $_2\pi_s + _2\pi_s$ cycloaddition constitutes a model for a photochemical aromatic reaction.

b) A thermal or photochemical TB $2\pi + 2\pi$ cycloaddition constitutes a model for a thermal or photochemical bicentric nonaromatic reaction.

c) A thermal $_2\pi_s + _2\pi_a$ cycloaddition constitutes a model for a thermal aromatic reaction, while a photochemical $_2\pi_s + _2\pi_a$ cycloaddition constitutes a model for a photochemical antiaromatic reaction. In general, the various P.E. surface aspects of thermal aromatic reactions resemble those of thermal nonaromatic reactions and the same is true of the corresponding (antiaromatic and nonaromatic, respectively) photochemical reactions.

The ideas presented in this section can be extended to any EE $\pi + \pi$ cycloaddition in the manner indicated below.

No. of Pi electrons	Stereochemical path	P.E. surface
4N	s + a	Fig. 28
4N	s + s	Fig. 27
4N + 2	s + a	Fig. 27
4N + 2	s + s	Fig. 28

Finally, it should be noted that nonionic $_2\pi_s + {}_2\pi_s$ cycloadditions are actually pseudoionic reactions. The term nonionic is used consistently to dramatize the fundamental difference of these reactions from ionic $_2\pi_s + {}_2\pi_s$ cycloadditions.

5. The Problem of Correlation Imposed Barriers

In the previous chapters we identified two extreme types of chemical reactions, i.e., reactions proceeding via thermal aromatic paths where interpacket interactions prevail and reactions proceeding via thermal antiaromatic paths where intrapacket interactions produce the final shapes of the P.E. surfaces. It is interesting to inquire how this picture developed on the basis of the LCFC method compares with the Woodward-Hoffmann approach which employs correlation diagrams in connection with wavefunctions which are delocalized over the entire reaction complex.

Consider two ethylenes interacting in a $_2\pi_s + _2\pi_s$ manner at a given intermolecular distance. The MO's of the reaction complex are shown in Fig. 40. Let us now write the determinental wavefunction for the ground configuration of the complex, i.e., the configuration which places two electrons in ϕ_1 and two electrons in ϕ_2.

$$\Psi^0 = |\phi_1(1)\,\bar{\phi}_1(2)\,\phi_2(3)\,\bar{\phi}_2(4)| \tag{5.1}$$

By reference to Fig. 40 we can substitute the LCAO form of ϕ_n. Dropping the normalization constants, obtain:

$$\Psi^0 \propto |(p_1 + p_2 + p_3 + p_4)(1)\,\overline{(p_1 + p_2 + p_3 + p_4)}(2)\,(p_1 + p_2 - p_3 - p_4)(3)$$
$$\overline{(p_1 + p_2 - p_3 - p_4)}(4)| \tag{5.2}$$

The above determinant can now be expanded in two different ways:

a) By making the substitution $p_1 + p_2 \propto \pi_{12}$ and $p_4 + p_3 \propto \pi_{43}$.

$$\Psi^0 \propto |(\pi_{12} + \pi_{43})(1)\,\overline{(\pi_{12} + \pi_{43})}(2)\,(\pi_{12} - \pi_{43})(3)\,\overline{(\pi_{12} - \pi_{43})}(4)| \tag{5.3}$$

This determinant contains three types of terms. Firstly, terms where the same spatial MO, π_{12} or π_{43}, makes a quadruple contribution, e.g., the term $|\pi_{12}(1)\,\bar{\pi}_{12}(2)\,\pi_{12}(3)\,\bar{\pi}_{12}(4)|$. Secondly, terms where the same spatial MO makes a triple contribution, e.g., the term $|\pi_{12}(1)\,\bar{\pi}_{43}(2)\,\pi_{43}(3)\,\bar{\pi}_{43}(4)|$. Finally, a term where the same spatial MO makes a double contribution, i.e., the term $|\pi_{12}(1)\,\bar{\pi}_{12}(2)\,\pi_{43}(3)\,\bar{\pi}_{43}(4)|$. The terms of the first and second types vanish and the determinant is reduced to a form which contains only one term of the third type. This is nothing else but the wavefunction of the reactant no bond configuration, $D_1 D_2$, describing the two ethylenic pi bonds.

$$\Psi_{D_1 D_2} \propto |\pi_{12}(1)\,\bar{\pi}_{12}(2)\,\pi_{43}(3)\,\bar{\pi}_{43}(4)| \tag{5.4}$$

Accordingly, we can say that Ψ^0 can be reduced to $\Psi_{D_1 D_2}$.

b) By making the substitution, $p_1 + p_4 \propto \sigma_{14}$, $p_2 + p_3 \propto \sigma_{23}$, $p_1 - p_4 \propto \sigma_{14}^*$ and $p_2 - p_3 \propto \sigma_{23}^*$.

$$\Psi^0 \propto |(\sigma_{14} + \sigma_{23})\,(1)\,\overline{(\sigma_{14} + \sigma_{23})}\,(2)\,(\sigma_{14}^* + \sigma_{23}^*)\,(3)\,\overline{(\sigma_{14}^* + \sigma_{23}^*)}\,(4)| \tag{5.5}$$

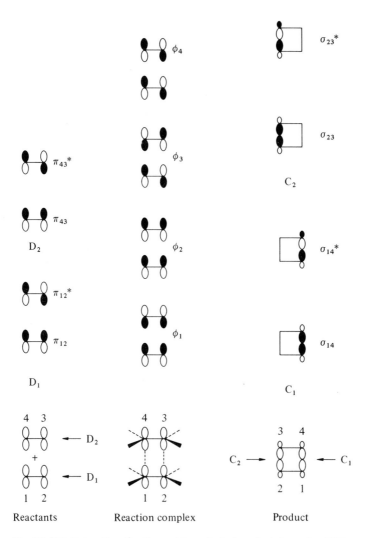

Fig. 40. MO designations for the problem of ethylene $2\pi + 2\pi$ cycloaddition

This determinant contains terms where the same spatial MO cannot appear more than twice. For example, consider the $|\sigma_{14}(1)\,\overline{\sigma_{14}}\,(2)\,\sigma_{14}^*(3)\,\overline{\sigma_{14}^*}\,|$ term. This is nothing else but the wavefunction of the product charge ditransfer configuration, $C_1^{-2}C_2^{+2}$, describing the two cyclobutane sigma bonds formed by the union of the two ethylenes. As a second example, consider the $|\sigma_{14}(1)\,\overline{\sigma_{14}}\,(2)\,\sigma_{23}^*(3)\,\overline{\sigma_{23}^*}\,(4)|$ term. This is the wavefunction of the product locally diexcited configuration, $C_1 C_2^{**}$, describing the two cyclobutane sigma bonds. Continuing in the same vein, we can identify all product configurations which arise from Ψ^0. These are $\Psi_{C_1^{**}C_2}$, $\Psi_{C_1 C_2^{**}}$, $\Psi_{C_1^{+2}C_2^{-2}}$, $\Psi_{C_1^{-2}C_2^{+2}}$, $\Psi_{C_1^{+*}C_2^-}$, $\Psi_{C_1^+ C_2^{-*}}$, $\Psi_{C_1^- C_2^{+*}}$, $\Psi_{C_1^{-*}C_2^+}$, and $\Psi_{C_1^* C_2^*}$.

The above analysis tells us that the no bond configurations of the reactants transforms into diexcited configurations of the product. The various transformations which can be obtained by this technique for the case of ethylene $2\pi + 2\pi$ cyclodimerization are shown in Table 11, which shows how a connection can be made between the LCFC approach and the delocalized approach most commonly employed today in computational quantum chemistry.

Table 11. Transformations of delocalized configurations to fragment configuration packets

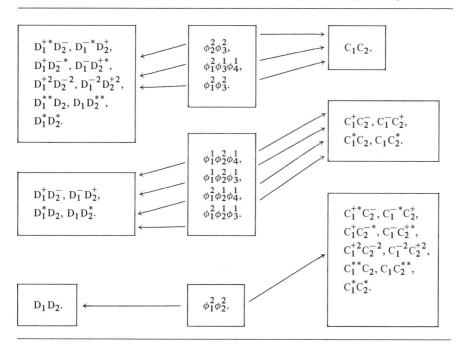

6. Reactivity Trends of Thermal Cycloadditions

6.1 Introduction

In this chapter, we compare specific predictions or general implications of the theoretical analysis of Chapt. 4 with the results of experiments. Since these experiments were carried out with aims other than testing the theory presented in this book, certain key pieces of information often are missing. Nonetheless, the available evidence is sufficient to delineate gross mechanistic trends.

6.2 Mechanisms of Stereochemical Nonretention in Cycloadditions

In Sect. 4.2, we constructed P.E. surfaces for cycloadditions by assuming that the two reactants approach each other in a symmetrical fashion involving equal overlap along both pairs of uniting termini. How may deviations from this ideal picture occur? Restricting our attention to thermal $2\pi + 2\pi$ cycloadditions, we distinguish the following situations.

a) In nonionic or ionic pericyclic $_2\pi_s + _2\pi_a$ cycloadditions, the ground surface results from $\Lambda_1 - \Lambda_2$ interpacket interactions. In general, there will be a tendency for maximizing multicentric spatial overlap since this will strengthen the $\Lambda_1 - \Lambda_2$ interpacket interactions.

b) In nonionic quasipericyclic $_2\pi_s + _2\pi_s$ cycloadditions, the ground surface is made up of two intersecting packets, the Λ_1 and the Λ_3 packets. The point of intersection amounts to a pseudocrossing. The transition state will be located in the vicinity of this intersection. Reduction of the thermal reaction barrier can be achieved if the geometry of approach simultaneously renders the $\Lambda_1 - \Lambda_2$ interpacket interactions nonzero without substantially reducing Λ_3 intrapacket interactions. Three possible geometries which meet this condition are shown below.

In summary, in the case of quasipericyclic cycloadditions, the two reaction partners may arrive in the vicinity of the transition state in a U-like, G-like, or R-like geometry. At this point, bond rotation can occur within one or both reaction partners. Depending on the size of the barrier, stereorandom or stereoselective product formation will occur.

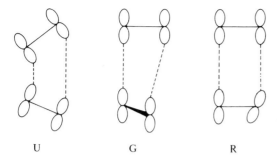

| U | G | R |

c) In ionic effectively pericyclic $_2\pi_s + _2\pi_s$ cycloadditions, the ground state surface is made up of the intersecting Λ_1, Λ_2, and Λ_3 packets. Barrier reduction can be accomplished by rendering $\Lambda_1 - \Lambda_2$ interpacket interactions nonzero without substantially reducing the strength of Λ_2 intrapacket interactions. Accordingly, as in the case of quasipericyclic reactions, the transition state may have the appearance of a U-like, G-like, or R-like structure and bond rotation could occur. Depending upon the size of the rotational barrier stereorandom or stereoselective product formation may arise.

In addition to the above multicentric mechanisms, a cycloaddition may occur via the TB mechanism. This will give rise to partial or complete loss of the stereochemical integrity of the reactants.

The rotational barrier in U-like, G-like, and R-like structures is predicted to be different in nonionic quasipericyclic and ionic effectively pericyclic reactions. The key arguments can be understood by means of a specific example. To this extent, consider U-like structures near the transition states of quasipericyclic and effectively pericyclic $_2\pi_s + _2\pi_s$ cycloaddition. As can be inferred from Fig. 27, U' is looser than U". U' resembles more two reactant molecules, both having nearly fully formed pi bonds, while U" resembles more a cisoid diradicaloid intermediate having no pi bonds. Hence, bond rotation in U" is expected to be more facile than in U'. Accordingly, stereorandomization is expected to be appreciable in the case of a quasipericyclic $_2\pi_s + _2\pi_s$ cycloaddition. By contrast, effectively pericyclic reactions may be as stereoselective as true pericyclic reactions.

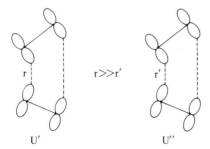

U'	U"
Effectively pericyclic reaction	Quasipericyclic reaction
Transition state	Transition state

The torsional motions which are expected to obtain in nonionic and ionic photo-cycloadditions can be discussed in a similar manner.

In general, the analysis presented above can be condensed in two rules:

a) A reaction complex which finds itself on an antipericyclic, antinonpericyclic or nonpericyclic path will tend to distort so that the total energy will decrease. If such a distortion occurs at long intermolecular distances, stereorandomization due to torsional motions within each reactant will be minimal. By contrast, if distortion occurs at short intermolecular distances, stereorandomization will be pronounced.

b) A reaction complex which finds itself on a pericyclic path will tend to remain undistorted.

6.3 Reactivity Trends of Thermal Nonionic $2\pi + 2\pi$ Cycloadditions

A $2\pi + 2\pi$ thermal nonionic cycloaddition can proceed in a pericyclic $_2\pi_s + _2\pi_a$, TB, or quasipericyclic $_2\pi_s + _2\pi_s$ manner. The favored $_2\pi_s + _2\pi_a$ pathway is sterically hindered and theory cannot predict the preferred stereochemical path definitively. However, each stereochemical path is associated with specific rate, regiochemical, topochemical etc., trends which can provide the basis for identifying the preferred stereochemistry of a $2\pi + 2\pi$ thermal nonionic cycloaddition. Accordingly, the experimental data should be analyzed *in toto* in an effort to determine whether a $_2\pi_s + _2\pi_s$, a TB, a $_2\pi_s + _2\pi_a$ mechanism, or any combination, can account for all or most known facts.

What are the stereochemical consequences of each of the three modes of approach of cycloaddends? In a $_2\pi_s + _2\pi_a$ approach, either cycloaddend can act as the antarafacial component, unless steric effects dictate one of the two possible modes. Competing $_2\pi_s + _2\pi_a$ mechanisms may lead to results expected from a nonstereoselective reaction. In the case of a TB approach, bond rotation can occur readily and such an approach is expected to be responsible for nonstereoselective reaction. Finally, in a $_2\pi_s + _2\pi_s$ approach, bond rotation may occur giving rise to nonstereoselective product formation. Clearly, any one of the three mechanisms may give rise to a low degree of stereoretention. In other words, *the stereochemical criterion for distinguishing $2\pi + 2\pi$ cycloaddition pathways is inadequate when applied by itself.*

As reaction polarity increases, the $_2\pi_s + _2\pi_s$ cycloaddition reaction system jumps onto a pericyclic path increasingly earlier along the reaction coordinate. *Thus, a reduction in stereorandomization is expected to accompany an increase in polarity if the operative mechanism is $_2\pi_s + _2\pi_s$.* No such effect is expected for competing $_2\pi_s + _2\pi_a$ mechanisms or the TB mechanism. Although no systematic study of the effect of polarity on stereoselectivity has ever been pursued, the available experimental evidence (Table 12) indicates that there is a gross trend: an increase in polarity is accompanied by an increase in stereoretention[1]. This trend is not smooth and the cycloadditions

---

1 All I—A values reported in this work have been computed using available experimental data and extrapolation techniques whenever necessary.

Table 12. Effect of increasing polarity on stereoretention in $2\pi + 2\pi$ retrocycloadditions and $2\pi + 2\pi$ cycloadditions

Reactants	$I_D - A_A$ (eV)	% Retention	References
	12.0	50	(1)
	12.0	40	(2)
	12.0	64	(3)
	9.8	50	(4)
	8.4	72	(5)
	12.0	75	(1)
	9.8	75	(4)
	8.4	95	(5)
	11.0	66	(6)

Table 12 (continued)

Reactants	$I_D - A_A$ (eV)	% Retention	References
MeO	9.5	88	(6)
+ ClCH=CHCl (cis)	11.6	68	(7)
+ MeCH=CHOEt (cis)	11.1	94	(8)
+ ClCH=CHCl (trans)	11.6	81	(7)
+ MeCH=CHOEt (trans)	11.1	79	(8)

(1) Cocks, A. T., Frey, H. M.: Chem. Commun. 1969, 458.
(2) Baldwin, J. E., Ford, P. W.: J. Amer. Chem. Soc. 91, 7192 (1969).
(3) Gerberich, H. R., Walters, W. D.: J. Amer. Chem. Soc. 83, 3935 (1961).
(4) Jones II, G., Williamson Jr., M. H.: J. Amer. Chem. Soc. 96, 5617 (1974).
(5) Jones II, G., Staires, J. C.: private communication.
(6) Paquette, L. A., Thompson, G. L.: J. Amer. Chem. Soc. 94, 7127 (1972).
(7) Jones Jr., M., Levin, R. H.: J. Amer. Chem. Soc. 91, 6411 (1969).
(8) Wasserman, H. H., Solodar, A. J., Keller, L. S.: Tetrahedron Let. 1968, 5597.

of benzyne and *trans* 1,2 disubstituted ethylenes are exceptional when compared to the corresponding cycloadditions of the cis isomers. It should be noted that, in comparing stereoselectivity, the same type of geometric isomer should be employed so that the steric bias for or against a $_2\pi_s + _2\pi_a$ process remains constant. Accordingly, only comparisons within each of the groups of Table 12 are valid.

We next turn our attention to HH versus HT regiochemistry of $2\pi + 2\pi$ cycloadditions. When two homodensic molecules are of the same type, i.e., both A type or both B type (see Table 6), $LU^D - LU^A$, $HO^D - HO^A$ and $HO^D - LU^A$ interactions all favor the same regiochemical preference, i.e., HH. As a result, *the regiochemistry of the $2\pi + 2\pi$ cycloaddition of two such homodensic molecules has absolutely no mechanistic significance* (Table 13). Ironically, *it is the regioselectivity of dimerization of homodensic molecules which has been invoked as evidence in support of*

Table 13. Head to head vs. Head to tail regioselectivity of $2\pi + 2\pi$ cycloadditions

Reactant type	Regioselectivity	References
A. Homodensic-homodensic		
2 CH_2=CHCN	HH	*(1)*
2 CH_2=CCl–CH=CH_2	HH	*(2)*
2 CH_2=CH–Ph	HH	*(3)*
2 CH_2=C(SMe)CN	HH	*(4)*
2 F_2C=CFCl	HH	*(5)*
2 Cl_2C=CF_2	HH	*(6)*
$\begin{array}{c} H_2C \\ \mid \\ H_2C \end{array}\!\!\!\!>\!\!C\!=\!CCl_2$	HH	*(7)*
B. Homodensic-heterodensic		
CF_2=CFCN + CH_2=CH–CH=CH_2	HH (CN, CHCH$_2$)	*(8)*
(Me)$_2$C=CHNMe$_2$ + CH_2=CHCN	HH (CN, NMe$_2$)	*(9)*
C. Heterodensic-heterodensic	–	

(1) Coyner, E. C., Hillman, W. S.: J. Amer. Chem. Soc. *71*, 324 (1949).
(2) Stewart Jr., C. A.: J. Amer. Chem. Soc. *93*, 4815 (1971).
(3) Brown, W. G.: J. Amer. Chem. Soc. *90*, 1916 (1968).
(4) Gunderman, K. D., Huchting, R.: Ber. 92, 415 (1959); Gunderman, K. D., Thomas, R.: Ber. *89*, 1263 (1956).
(5) Lacher, J. R., Tompkin, G. W., Park, J. D.: J. Amer. Chem. Soc. *74*, 1693 (1952).
(6) Henne, A. L., Ruh, R. P.: J. Amer. Chem. Soc. *69*, 279 (1947).
(7) Dolbier Jr., W. R., Lomas, D., Tarrant, P.: J. Amer. Chem. Soc. *90*, 3594 (1968).
(8) Roberts, J. D., Sharts, C. M.: Org. Reactions *12*, 1 (1962).
(9) Brannock, K. C., Bell, A., Burpitt, R. D., Kelley, C. A.: J. Org. Chem. *29*, 801 (1964).

the "diradical" hypothesis [31, 32] *so frequently advanced by organic chemists.* This hypothesis will be commented upon later.

The regiochemistry of a nonionic $2\pi + 2\pi$ cycloaddition can have mechanistic significance only if the reaction involves two heterodensic molecules, such that the $HO^D - LU^A$ interaction regiochemical consequences are different from those of the $HO^D - HO^A$ and $LU^D - LU^A$ interactions. However, even in such cases $HO^D - LU^A$ control is consistent with any of the $_2\pi_s + _2\pi_s$, TB, or $_2\pi_s + _2\pi_a$ mechanisms. Only if $HO^D - HO^A$ or $LU^D - LU^A$ control which is different from $HO^D - LU^A$ control obtains can a case for $_2\pi_s + _2\pi_s$ mechanism be made. Unfortunately, we have not been able to locate any studies of nonionic $2\pi + 2\pi$ cycloadditions of two heterodensic molecules.

Whereas $2\pi + 2\pi$ cycloadditions of two homodensic molecules of the same type are mechanistically unenlightening insofar as HH vs. HT regioselectivity is concerned,

$2\pi + 2\pi$ cycloadditions of one homodensic and one heterodensic molecule are ambiguous. In such cases one HO—LU interaction favors HH and the other HT regioselectivity. Also, HO—HO interaction favors one of the two regioisomers and LU—LU interaction the other.

Next, we consider the possibility of syn regioselectivity in $2\pi + 2\pi$ nonionic cycloadditions. This type of regioselectivity is not consistent with either a $_2\pi_s + _2\pi_a$ or a TB mechanism. It can be consistent with a $_2\pi_s + _2\pi_s$ mechanism *if the transition state is not sufficiently distorted so that syn secondary spatial orbital overlap is poor.* In short, syn regioselectivity in $2\pi + 2\pi$ cyclodimerizations points to a $_2\pi_s + _2\pi_s$ mechanism.

Experimentally, little effort has been spent in probing syn regioselectivity in appropriately designed systems. Table 14 shows that there are cases where syn regioselectivity does obtain and others where it does not.

Table 14. Syn versus anti regioselectivity of thermal $2\pi + 2\pi$ Head to Head Cycloadditions

Reactants	Syn/Anti ratio	References
$2\ Ph–CH=CH_2$	0.45	*(1)*
2	Large	*(2)*
$2\ CF_2=CFCl$	5.00	*(3)*
$2\ CH_2=CHCN$	0.67	*(4)*
$2\ CH_2=CCl–CH=CH_2$	0.28	*(5)*

(1) Brown, W. G.: J. Amer. Chem. Soc. *90,* 1916 (1968).
(2) Schröder, G., Martin, W.: Angew. Chem. Int. Edit. Engl. *5,* 130 (1966).
(3) Lacher, J. R., Tompkins, G. W., Park, J. D.: J. Amer. Chem. Soc. *74,* 1693 (1952).
(4) Bellǔs, D.: private communication; Coyner, E. C., Hillman, W. S.: J. Amer. Chem. Soc. *71,* 324 (1949).
(5) Stewart Jr., C. A.: J. Amer. Chem. Soc. *93,* 4815 (1971).

The effect of reaction polarity on the rates of nonionic $2\pi + 2\pi$ cycloadditions has not been studied extensively. However, the available data seem to suggest that increased polarity leads to faster reaction, *regardless of mechanism,* as expected. Typical data are shown in Table 15.

The selectivity-polarity relationship has not been probed yet. This could be a potentially fruitful area of research.

Solvent effects on reaction rates are expected to be very small in nonionic $2\pi + 2\pi$ cycloadditions because the transition state wavefunction is dominated by the no bond and triplet diexcited type configurations. In agreement with this prediction, it is found that an increase in solvent polarity leads to small rate acceleration (Table 16).

The experimental results considered *in toto* suggest that many thermal nonionic $2\pi + 2\pi$ cycloadditions may be true antiaromatic reactions. Thus, only the $_2\pi_s + _2\pi_s$ mechanism can successfully account for the polarity-stereoselectivity relationship and the syn regioselectivity of certain reactions. However, while many nonionic $2\pi + 2\pi$ cycloadditions may be antiaromatic processes, the possibility of competing $_2\pi_s + _2\pi_s$, TB and $_2\pi_s + _2\pi_a$ paths having thresholds separated by small energy gaps should also be considered.

Table 15. Activation Energies for $2\pi + 2\pi$ cycloreversions

Reactant	E_a(kcal/mole)	$I_D - A_A$(eV)	References
	62.5	12.3	(1)
	62.0	11.5	(2)
	51.0	11.0	(3)
CHO	53.3	10.5	(4)
H_3C O O Ph	22.9	9.3	(5)

(1) Genaux, C. T., Kern, F., Walters, W. D.: J. Amer. Chem. Soc. 75, 6196 (1953).
(2) Pataracchia, A. F., Walters, W. D.: J. Phys. Chem. 68, 3894 (1964).
(3) Ellis, R. J., Frey, H. M.: Trans. Far. Soc. 59, 2076 (1963).
(4) Roquitte, B. C., Walters, W. D.: J. Amer. Chem. Soc. 84, 4049 (1962).
(5) Richardson, W. H., Yelvington, M. B., O'Neal, H. E.: J. Amer. Chem. Soc. 94, 1619 (1972).

Table 16. Solvent effects on $2\pi + 2\pi$ and $4\pi + 2\pi$ thermal cycloadditions

Reaction	$\dfrac{k(MeCN)}{k(Cyclohexane)}$	$\dfrac{k(MeCN)}{k(CCl_4)}$	$\dfrac{k(EtOH)}{k(Benzene)}$	References
A. Nonionic $2\pi + 2\pi$				
CH$_2$=CHOn-Bu + (Ph)$_2$C=C=O	160			(1)
Me$_2$C=C=CH$_2$ + CH$_2$=CH−CN	<2[a]			(2)
CH$_2$=CH−CH=CH$_2$ + Cl$_2$C=CF$_2$	2.75[b]			(3)
B. Ionic $2\pi + 2\pi$				
p-AnCH=CH$_2$ + TCNE		63,000	75[c]	(4, 5)
p-AnCH=CHMe (trans) + TCNE	29,000			(5)
(Me)$_2$C=CHOEt + TCNE	10,800			(5)
CH$_2$=CHO*n*−Bu + TCNE	2,600			(5)
 + TCNE		17,000		(5)
MeCH=CHOMe(cis) + TCNE		10,330		(6)
MeCH=CHOMe(trans) + TCNE		8,869		(6)
C. Nonionic $4\pi + 2\pi$				
 + (Ph)$_2$C=C=O	26			(7)

Table 16 (continued)

Reaction	k(MeCN)/k(Cyclohexane)	k(MeCN)/k(CCl₄)	k(EtOH)/k(Benzene)	References
	≤10			(7)
			2.8	(8)
			4.6	(9)
			1.47[d]	(10)

[a] k(DMSO)/k(cyclohexane).
[b] k(MeNO₂)/k(hexane).
[c] k(EtOH)/k(Toluene)
[d] k(MeNO₂)/k(benzene).

(1) Huisgen, R., Feiler, L. A., Otto, P.: Tetrahedron Let. 1968, 4485.
(2) Kiefer, E., Okamura, M. Y.: J. Amer. Chem. Soc. 90, 4187 (1968).
(3) See [31].
(4) Williams, J. K., Wiley, D. W., McKusick, B. C.: J. Amer. Chem. Soc. 84, 2210 (1962).
(5) See [36].
(6) See [37].
(7) See [40].
(8) Kaufmann, H., Wassermann, A.: J. Chem. Soc. 1939, 870.
(9) Wassermann, A.: J. Chem. Soc., 1935, 828.
(10) Wong, K. F., Eckert, C. A.: Trans. Far. Soc. 66, 2313 (1970).

113

6.4 Reactivity Trends of Thermal Ionic $2\pi + 2\pi$ Cycloadditions

$2\pi + 2\pi$ thermal ionic cycloadditions can proceed in a pericyclic $_2\pi_s + _2\pi_a$ manner via an N^* intermediate, a TB manner, or, an effectively pericyclic $_2\pi_s + _2\pi_s$ manner via an M intermediate. *All three stereochemical paths are predicted to involve inter-mediates having dipolar character.* Electronically, the energy separation of the $_2\pi_s + _2\pi_a$ and $_2\pi_s + _2\pi_s$ paths is small when the DA and D^+A^- diabatic surfaces cross at large intermolecular distance. Furthermore, the former pathway is made nearly impossible by virtue of the substituents borne by the two olefins, i.e., steric effects are prohibitive.

The energy separation of the $_2\pi_s + _2\pi_s$ and TB paths is also small on *electronic* grounds. In addition, coulombic effects, i.e., C in equation 4.16, clearly favor the former over the latter. Once the M intermediate is formed, solvent effects may augment an already appreciable barrier (due to pericyclic bonding) for isomerization to the TB intermediate. Thus, the effectively pericyclic $_2\pi_s + _2\pi_s$ path is expected to be followed. Indeed, experimental evidence, summarized in Table 17, indicates that ionic $2\pi + 2\pi$ cycloadditions are highly stereoselective, or, even stereospecific, in a $_2\pi_s + _2\pi_s$ sense.

Table 17. Stereochemistry of polar cycloadditions

Donor	Acceptor	% Formal 2s + 2s	References
Et–S ___ Me	COOMe \ N=N \ COOMe	100%	(1)
	NC CN \ / NC CN	>90%	(2)
	NC CN \ / NC CN	>90%	(2)
Me, OMe (aryl)	NC CN \ / NC CN	85–90% (PhH) 72% (CH_2Cl_2) 51% (CH_3CN)	(3)

Table 17 (continued)

Donor	Acceptor	% Formal 2s + 2s	References
Me / OMe (styrene derivative)	NC, CN / NC, CN	100% (PhH, CH$_2$Cl$_2$, CH$_3$CN)	(3)
Me / Me (cyclohexene)	NC, CN / NC, CN	100%	(3)
Me / Me (cyclohexene)	NC, CN / NC, CN	100%	(3)
Me, Me / Me / Me	NC, CN / NC, CN	100%	(3)
Me, Me / Me	NC, CN / NC, CN	72.3%	(3)
Me — OMe	NC, CN / NC, CN	95% (PhH) 94% (CH$_2$Cl$_2$) 91% ((CH$_3$)$_2$CO) 84% (CH$_3$CN)	(4)
OMe / Me	NC, CN / NC, CN	97% (PhH) 95% (CH$_2$Cl$_2$) 91% (CH$_3$COOCH$_2$CH$_3$) 80% (CH$_3$CN)	(4)

Table 17 (continued)

Donor	Acceptor	% Formal 2s + 2s	References
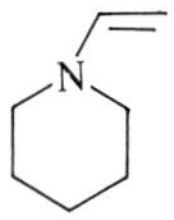 (N-vinyl piperidine)	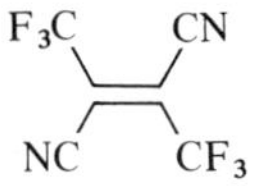 F₃C, CN / NC, CF₃	100%	(5)
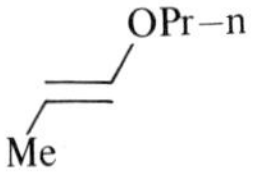 OPr-n / Me	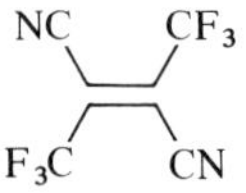 NC, CF₃ / F₃C, CN	100%	(5)
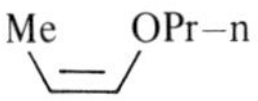 Me, OPr-n	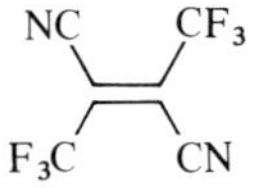 NC, CF₃ / F₃C, CN	46.2%	(5)
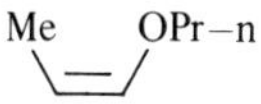 Me, OPr-n	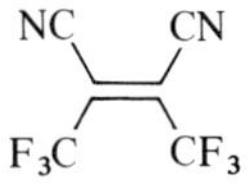 NC, CN / F₃C, CF₃	59.5%	(5)
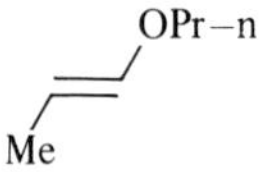 OPr-n / Me	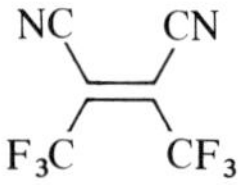 NC, CN / F₃C, CF₃	100%	(5)
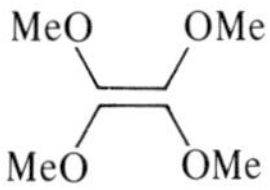 MeO, OMe / MeO, OMe	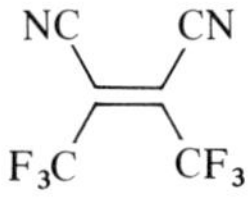 NC, CN / F₃C, CF₃	100%	(6)
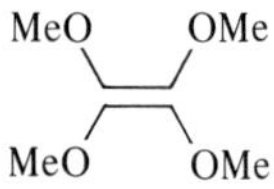 MeO, OMe / MeO, OMe	 NC, CN	100%	(7)
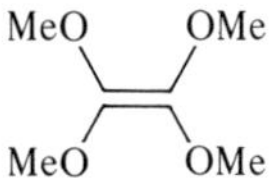 MeO, OMe / MeO, OMe	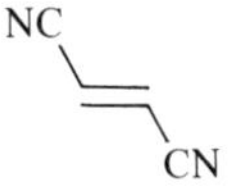 NC, CN	100%	(7)
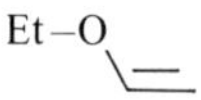 Et–O	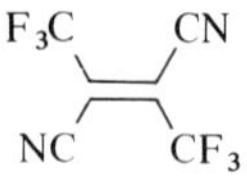 F₃C, CN / NC, CF₃	100%	(5)

116

Table 17 (continued)

Donor	Acceptor	% Formal 2s + 2s	References
Et–O ⟍⟍═	NC ⟍ CN / F₃C ⟍ CF₃	100%	(5)
tBu–S ⟍⟍═	F₃C ⟍ CN / NC ⟍ CF₃	100%	(5)
tBu–S ⟍⟍═	NC ⟍ CN / F₃C ⟍ CF₃	100%	(5)
⟨ring⟩ OMe	F₃C ⟍ CN / NC ⟍ CF₃	100%	(5)
⟨ring⟩ OMe	NC ⟍ CN / F₃C ⟍ CF₃	100%	(5)

(1) Firl, J., Sommer, S.: Tetrahedron Let. *1972*, 4713.
(2) Nishida, S., Moritani, I., Teraji, T.: J. Org. Chem. *38*, 1878 (1973).
(3) Bartlett, P. D.: Quart. Revs. *24*, 473 (1970).
(4) Huisgen, R., Steiner, G.: J. Amer. Chem. Soc. *95*, 5054 (1973).
(5) Proskow, S., Simmons, H. E., Cairns, T. L.: J. Amer. Chem. Soc. *88*, 5254 (1966).
(6) Hoffmann, R. W.: Angew. Chem. Internat. Ed. Engl. *7*, 754 (1968).
(7) Hoffmann, R. W., Bressel, U., Gehlhaus, J., Hauser, H.: Chem. Ber. *104*, 873 (1971).

The regiochemistry of the $2\pi + 2\pi$ thermal ionic cycloadditions is predicted to be the one which maximizes the $HO^D - LU^A$ interaction regardless of which stereochemical path is followed. Hence, *the regiochemical criterion cannot provide any valuable mechanistic information in the case of ionic $2\pi + 2\pi$ cycloadditions.*

Ionic $2\pi + 2\pi$ cycloadditions, like most ionic reactions, are expected to be extremely fast and that is precisely what the experimental evidence indicates. Indeed, some of these reactions are so facile that occur with appreciable rate even at temperatures below $0°$ [33]. Accordingly, they constitute an important class of reactions having great synthetic utility.

The effect of substituents on the rates of ionic $2\pi + 2\pi$ cycloadditions is not only highly interesting but is also highly suggestive of a Hückel antiaromatic path. Since the transition state of an ionic $_2\pi_s + _2\pi_s$ cycloaddition depends directly upon the $HO^D - LU^A$ interaction, reactant symmetry becomes crucial, i.e., a pronounced effect of unsymmetrical reactant substitution on reaction rate is a strong indication of the operation of the $_2\pi_s + _2\pi_s$ mechanism. Such an effect is observed in ionic $2\pi + 2\pi$ cycloadditions as the data of Tables 18 and 19 demonstrate.

The contrast between the selectivity of thermal nonionic and thermal ionic $2\pi + 2\pi$ cycloadditions is instructive. In nonionic reactions, a direct comparison of aromatic and antiaromatic paths cannot be made. However, in ionic reactions, such a comparison is possible; it is expected that increased polarity will lead to increasingly looser transition states and smaller energy separation of aromatic and antiaromatic paths. As a result, increasing the donor or acceptor ability of one reaction partner is expected to shift the electronically imposed preference for an aromatic path to a preference for an antiaromatic path favored by steric and/or electrostatic effects. A typical example is given below [34]:

Finally, an extension of the key concepts developed in this and previous chapters to superionic $2\pi + 2\pi$ cycloadditions is interesting: when one reactant constitutes an extremely strong donor and the other an extremely strong acceptor, the minimum of the D^+A^- diabatic surface may lie underneath the reactant energy level and the $D^{+2}A^{-2}$ diabatic surface may become the lowest energy diexcited diabatic surface. As a result, there could be very weak mixing of D^+A^- with higher lying diabatic surfaces near its minimum and the same could be true for the $D^{+2}A^{-2}$ diabatic surface. Such reactions will proceed with initial formation of two radical ions $D^{+\cdot}$ and $A^{-\cdot}$ followed by formation of two closed shell ions D^{++} and $A^=$. Indeed, such entities have been observed when a super donor and a super acceptor olefin were brought in the company of each other. For example, tetrakis-(dimethylamino)-ethylene, which has an ionization potential very close to that of sodium, and tetracyanoethylene react as shown below [35].

118

Solvent effects on the rates of ionic $_2\pi_s + {_2}\pi_s$ cycloadditions are predicted to be very large regardless of mechanistic type (A or B). This is due to the fact that in a type A mechanism the transition state wavefunction is made up of equal weights of no bond and charge transfer configurations, while in a type B mechanism the charge transfer contribution is still large and, in fact, greater than the no bond contribution. In addition, the asymmetry of the transition state accentuates ist dipolar character. Experimental investigations have demonstrated that increased solvent polarity leads to a pronounced rate increase (Table 16).

6.5 The $2\pi + 2\pi$ Ionic Cycloaddition Problem

The currently accepted mechanism of thermal ionic $2\pi + 2\pi$ cycloadditions is shown below.

Cisoid
biradicaloid (CB)

This mechanism is favored by experimentalists such as Huisgen [36–39], Gompper [40], Bartlett [31], and others [41], has been assumed as valid by certain theoreticians [42] and will be labelled the CB mechanism because its key feature is the formation of a CB structure involving no 1–4 bonding. By contrast, we have argued that most ionic $2\pi + 2\pi$ cycloadditions are effectively pericyclic $_2\pi_s + {_2}\pi_s$ reactions involving formation of an M intermediate which is pericyclically bonded.

Before we turn our attention to the key differentiating data, we enumerate four types of mechanistic evidence which can be accounted for by *either mechanism:*

a) Trapping of intermediate [38].
b) Rate response to polar solvents (see Table 16).
c) The sign and magnitude of $\Delta V^\dagger$ and $\Delta S^\dagger$ [39, 40].
d) Measurable isomerization of the cycloaddends [31, 36].

The differentiating pieces of evidence are as follows:

a) Ionic $2\pi + 2\pi$ cycloadditions are highly stereoselective or even stereospecific. This observation is incompatible with the CB mechanism because a "configuration holding" mechanism in the presumably nonpericyclic intermediate is necessary to account for the observed stereoselectivity. This mechanism was proposed to be coulombic attraction between the ends of the dipole, something which does not make sense because, in the point charge approximation, rotation in the intermediate could occur without impairing coulombic attraction.

At this point, 9 emphasize again [43] that the $_2\pi_s + {_2}\pi_s$ path ends up being preferred because of the following reasons:

1. The *electronic* separation of the various paths is small due to the early crossing of DA and D^+A^-. A consideration of steric and coulombic effects suggests preference for the $_2\pi_s + _2\pi_s$ path.

2. The intrinsic barrier involved in the M → TB transformation is appreciable due to pericyclic bonding. It may be further enhanced by solvent effects.

Proponents of the CB mechanism have often cited inappropriate data in attempts to support their ideas. Thus, the observed stereospecific addition of trans anethole to TCNE contrasts with the nonstereoselective addition of the cis isomer to the same acceptor [31]. This could be taken as evidence for a freely rotating intermediate. However, a $_2\pi_s + _2\pi_s$ approach is hindered in the case of the cis isomer because in this molecule the phenyl group is rotated by ca. 55° out of plane [44]. Thus, this system is biased towards a TB mechanism.

b) Trapping of the intermediate is stereospecific in a manner consistent with the presence of 1–4 bonding [38]. Thus, as illustrated below, the intermediate formed by cis-propenyl methyl ether and TCNE can be trapped at 0 °C with 95% stereoselectivity.

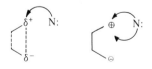

X = OEt   Y = OMe   80%
X = OMe   Y = OEt   4%

If the intermediate had no 1–4 pericyclic bonding, the alcohol would add from either of two possible directions. Since addition is only from the "outside" [38], there must be some 1,4 bonding.

M intermediate   CB intermediate

c) The effect of unsymmetrical substitution on the rates of ionic $2\pi + 2\pi$ cycloadditions. This has already been discussed. One may still argue that a CB mechanism is consistent with the observations because both CB and $_2\pi_s + _2\pi_s$ mechanisms lead to a greater $\langle \mathrm{DA} | \hat{P} | \mathrm{D^+A^-} \rangle$ matrix element in the unsymmetrical than the symmetrical case. However, the *difference* is *smaller* in the case of the CB mechanism. This can be understood from inspection of the computational data shown below, where k is a composite energy constant. The change in the $\mathrm{HO}^D - \mathrm{LU}^A$ matrix element in going from a symmetrical to an unsymmetrical donor olefin is much more pronounced in the case of the $_2\pi_s + _2\pi_s$ mechanism.

Table 18

Model system	$\langle \mathrm{HO}^D | \hat{P} | \mathrm{LU}^A \rangle$	
	CB mechanism	$_2\pi_s + _2\pi_s$ mechanism
HOCH=CHOH + TCNE	0.502 k	0.0
CH_3CH=CHOH + TCNE	0.586 k	0.110 k
Δ	0.084 k	0.110 k

In the author's opinion, a CB mechanism cannot account, *at least as well as the* $_2\pi_s + _2\pi_s$ *mechanism,* for the very pronounced difference in reactivity between the symmetrical and unsymmetrical olefins.

Additional interesting experimental results supporting the $_2\pi_s + _2\pi_s$ mechanism can be found in the literature. For example, consider the thermal cycloaddition shown below. Now, substitution of H_1 and H_2 by *weak* donors, such as methyls, will not have a pronounced effect on

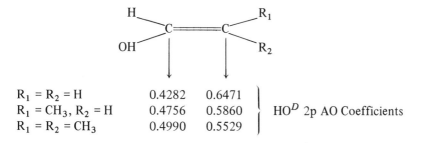

the electron donating ability but will tend to equalize the HO^D 2p AO coefficients of the vinylic carbons, and, thus, drastically reduce the $\langle \mathrm{DA} | \hat{P} | \mathrm{D^+A^-} \rangle$ matrix element. Results of CNDO/2 calculations, which show how this type of substitution tends to "symmetrize" the HO electron density, are shown below:

$R_1 = R_2 = H$	0.4282	0.6471	
$R_1 = CH_3, R_2 = H$	0.4756	0.5860	HO^D 2p AO Coefficients
$R_1 = R_2 = CH_3$	0.4990	0.5529	

If polar cycloadditions are $_2\pi_s + {}_2\pi_s$ effectively pericyclic reactions, substitution of H_1 and H_2 by alkyl groups may *decrease* the reaction rate. This has been found experimentally, as shown in Table 19. In the author's opinion, these results cannot be explained as well on the basis of the CB mechanism.

Huisgen has attempted to provide a rationalization of the aforementioned reactivity trends on the basis of the CB mechanism by invoking several effects. It is interesting to examine those effects in some detail:

a) In order to account for the greater reactivity of 1-butenyl ether relative to *cis* and *trans* 1, 2 diethoxyethylene, Huisgen argued that "... donor activities of the two ethoxy groups seem to impair each other". However, this suggestion is incorrect since any CHR=CHR, where R is a pi donor, is a better donor than CHR=CH$_2$, as the following ionization potential data reveal.

Molecule	Ionization potential (eV)
MeOCH=CH$_2$	8.93 [45]
MeOCH=CHOMe, *cis*	7.97 [46]
MeOCH=CHOMe, *trans*	8.04 [46]
ClCH=CH$_2$	10.18 [47]
ClCH=CHCl, *cis*	9.83 [47]
ClCH=CHCl, *trans*	9.81 [47]

b) In order to account for the greater reaction rate of 1-butenyl ether relative to the 1,2 diethoxyethylenes toward TCNE, Huisgen invoked steric effects. However, if steric effects were of paramount importance, a TB mechanism would be more favorable than the CB mechanism. This possibility is not supported by the experimental evidence, i.e., a TB mechanism cannot account for the reaction stereoselectivity and the stereoselective trapping of the intermediate.

Table 19. Unsymmetrical substitution effects in thermal ionic $2\pi + 2\pi$ cycloadditions[a]

	Temperature	Reaction time
	Room	1–2 sec.
	Room	1.4×10^5 sec.

Y = CN

[a]   Nishida, S., Moritani, I., Teraji, T.: J. Org. Chem. *38*, 1878 (1973).

Although the experimental evidence cited above suggests that the reactions of TCNE or $(CN)_2C=CH(CN)$ with good electron donors are effectively pericyclic $_2\pi_s + _2\pi_s$ ionic reactions, the alternative of a pseudoionic process cannot be dismissed presently. A mechanism of the latter type could also account satisfactorily for all the reactivity trends rationalized by the ionic mechanism.

6.6 Reactivity Trends of Thermal Nonionic $4\pi + 2\pi$ Cycloadditions

Thermal $4\pi + 2\pi$ cycloadditions are predicted to occur via a $_4\pi_s + _2\pi_s$ mechanism. The pertinent qualitative P.E. surface is analogous to that shown in Fig. 28 a. Experimental evidence in support of this prediction has been well surveyed and needs no further discussion [48, 49].

The preferred regiochemistry of $_4\pi_s + _2\pi_s$ cycloadditions is predicted to be the one which maximizes the $HO^D - LU^A$ interaction. Good agreement between theory and experiment is found in substantially polar though nonionic reactions [49, 26]. The regioselectivity of highly nonpolar nonionic $4\pi + 2\pi$ cycloadditions has not been studied systematically. Compilations of experimental results regarding the regioselectivity of nonionic $4\pi + 2\pi$ cycloadditions can be found in various articles and reviews [49, 50]. Secondary overlap effects give rise to endo over exo preference unless steric effects mitigate otherwise [4a].

The effect of reaction polarity has been extensively investigated. In general, increased reaction polarity leads to increased reaction rates. When polarity differences are not large, the MO electron density "dilution" effect is manifested reversing the normally expected reactivity order. A typical example is given below [49][1].

$k_{rel} = \sim 23$

$k_{rel} = 1$

---

1 Replacement of maleic anhydride by TCNE, may render the reactions with butadiene and 1,4-diphenylbutadiene ionic. In such a case, the interaction matrix element difference is small due to the early crossing of DA and D^+A^- and the relative rate may now be controlled by polarity, i.e., 1,4-diphenylbutadiene may now react faster. This turnover has been observed experimentally (see [49]).

The effect of catalysts on the rate of $4\pi + 2\pi$ cycloadditions has been known for quite some time. Coordination of an appropriate catalyst with the dienophile can increase reaction polarity and lead to enhanced reaction rates [49].

Interestingly, nonionic $4\pi + 2\pi$ cycloadditions of extremely low polarity, like all thermal aromatic reactions of this type, are predicted to go through a tight (late), high energy transition state. As a result, steric effects may well become more important than electronic effects and they may dictate an alternative reaction path, especially if one or both reactants bear bulky substituents. In this connection, it has been observed that the Diels-Alder reaction of hexachlorocyclopentadiene and trans-1,2-dichloroethylene or fumaronitrile is nonstereospecific, while the Diels-Alder reaction of hexachlorocyclopentadiene and trans-2-butene is stereospecific [51]. The former combinations represent nonionic reactions of very low polarity while the latter combination a nonionic reaction of moderate polarity.

The dependence of $4\pi + 2\pi$ cycloaddition regioselectivity upon polarity has not been studied systematically. However, catalysts have been known to increase the regioselectivity of certain $4\pi + 2\pi$ cycloadditions. This is explicable in terms of our analysis, if it is assumed that the catalyst increases the reaction polarity, as evidenced by rate enhancement, without having an adverse effect on the electron density of HO^D and LU^A. An example is given below [52] and others can be found in the literature [53].

	Ph COOMe + COOMe	Ph COOMe
Uncatalyzed:	80%	20%
Catalyzed:	97%	3%

Solvent effects on the rates of nonionic $4\pi + 2\pi$ cycloadditions are expected and found to be small due to the fact that the transition state wavefunction has principal no bond character (Table 16).

6.7 Reactivity Trends of Thermal Ionic $4\pi + 2\pi$ Cycloadditions

Thermal ionic $4\pi + 2\pi$ cycloadditions are predicted to occur via a $_4\pi_s + _2\pi_s$ mechanism. The pertinent qualitative P.E. surface is analogous to that shown in Fig. 28. An N* intermediate is predicted and the consequences of such a molecular species on the reaction coordinate can be very interesting. For example, such an intermediate may react chemically with solvent or solute. The incursion of intermediates in ionic $4\pi + 2\pi$ cycloadditions has been recognized in several cases [40]. A possible example is given below [54], i.e., the reaction is most likely ionic on the basis of our estimates, yet very close to the borderline region so that some uncertainty still exists.

124

In thermal ionic $_2\pi_s + {_2}\pi_s$ cycloadditions, where the M intermediate involves pericyclic bonding, as well as in thermal ionic $_4\pi_s + {_2}\pi_s$ cycloadditions, where the N* intermediate also involves pericyclic bonding, pronounced asymmetry of the reactants may produce strong bonding along one pair of termini and weak along the other. Hence, such intermediates may have low rotational barriers and give rise to stereorandom cycloadduct formation. This possibility has not yet been investigated systematically.

The regioselectivity of ionic $4\pi + 2\pi$ cycloadditions is extremely interesting because the transition states of these reactions are loose and electrostatic effects may dominate "orbital effects". No systematic studies have yet been reported. This is also true of solvent effects on the rates of ionic $4\pi + 2\pi$ cycloadditions which are expected to be large regardless of mechanistic type (A or B).

Finally, one imaginative kinetic study has produced a possible example of a super-ionic $4\pi + 2\pi$ thermal cycloaddition [55][1].

---

1 The qualitative P.E. surfaces suggest that aromatic and nonaromatic ionic reactions must be interpreted assuming two distinct intermediates, aside from reactants and products, i.e., an encounter complex and an M or N* intermediate. The possibility of collapse of, e.g., N* and product to a single minimum in certain cases is left open.

7. Reactivity Trends of Singlet Photochemical Cycloadditions

7.1 Introduction

The analysis of photochemical cycloadditions involves as a first task the identification of the excited diabatic surface manifold. Fortunately, in most organic photoreactions, the lowest excited diabatic surface at infinite intermolecular distance is of the DA* or D*A type. The second task is the classification of a photocycloaddition according to the interrelationship of the DA, D^+A^-, and DA* (or D*A) diabatic surfaces by reference to the definitions of section 3.6 and the appropriate data.

A thorough analysis of various photocycloadditions has been carried out in our laboratories and a summary of the results is presented below. A proposed mechanism cannot be *adequately* supported by experimental evidence unless concordant stereochemical *and* regiochemical data are available. Unfortunately, this is often not the case. In most photocycloadditions, only *partial* support of a proposed mechanism could be obtained from the available literature data.

7.2 The Chorochemistry of Singlet $2\pi + 2\pi$ Photocycloadditions

Nonionic $2\pi + 2\pi$ photocycloadditions initiated by $\pi\pi^*$ excitation are expected to be $_2\pi_s + _2\pi_s$ stereoselective. Furthermore, the regioselectivity of these reactions is expected to be consistent with maximization of $HO^D - HO^A$ (DA* lower in energy than D*A) interactions. The existing data provide adequate (e.g., entry C6 in Table 20) and partial (e.g., entries C3, C4, C5, and C8 in Table 20) support of these predictions.

Nonionic $2\pi + 2\pi$ photocycloadditions initiated by $n\pi^*$ excitation are expected to be $_2\pi_s + _2\pi_s$ stereoselective. On the other hand, the regioselectivity of these reactions depends upon polarity. In nonpolar cases, the preferred regiochemistry is $LU^D - LU^A$ controlled; in polar cases, it is $n-HO^D$ controlled. Again, the experimental data seem to be consistent with these predictions (e.g., entries A4, A6, and A7 in Table 20).

126

Table 20. The chorochemistry of photochemical cycloadditions

Reactants	Products	$I_D - A_A$ (eV)	References

A. Singlet $n\pi^*$ Photocycloadditions

Reactants	Products	$I_D - A_A$ (eV)	References
1.		7.3	(1)
2.		7.8	(2)
3.		8.0	(3)

128

Table 20 (continued)

Reactants	Products	$I_D - A_A$ (eV)	References
4. 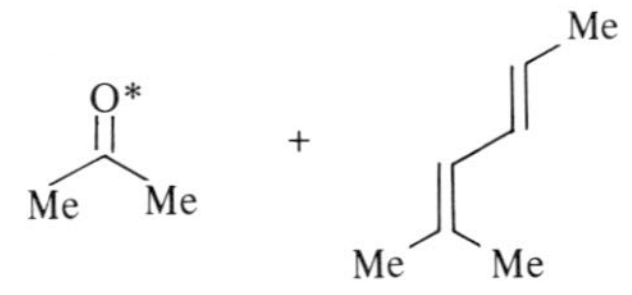	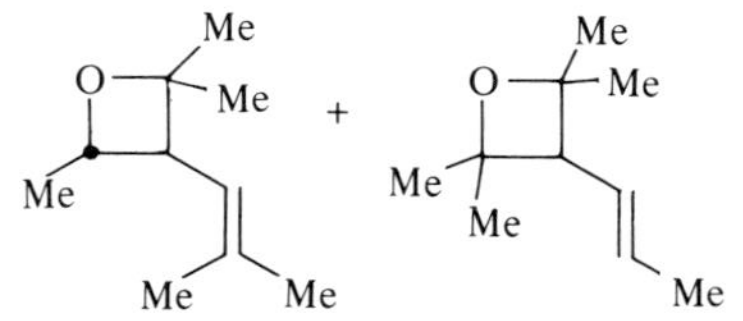 50% 50%	8.0	(4)
5. 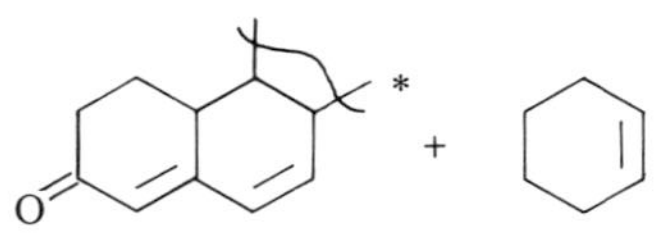	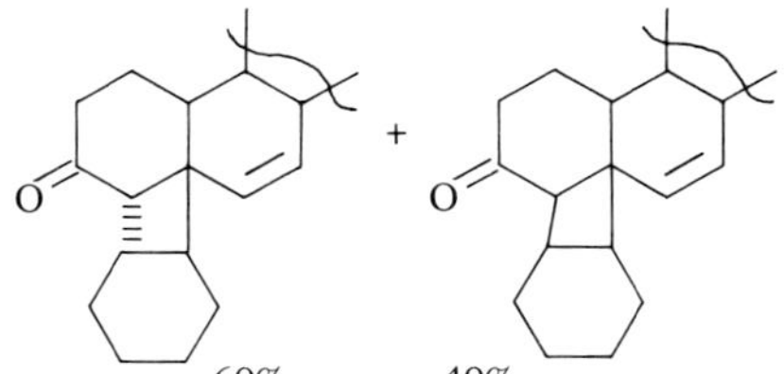 60% 40%	8.3	(2)
6. 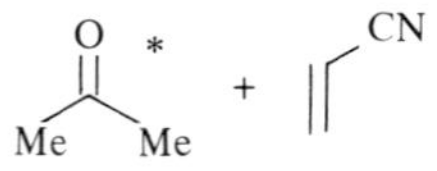	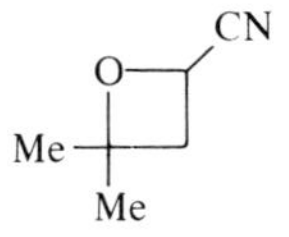	10.9	(5)

Table 20 (continued)

Reactants	Products	$I_D - A_A$ (eV)	References
7. + NCHC=CHCN	cis: trans:	11.2	(6)
		11.2	(6)
B. Triplet $n\pi^*$ Photocycloadditions			
1.	60% + 40%	6.3	(7)
2.	+	6.8	(8)

Table 20 (continued)

Reactants	Products	$I_D - A_A$ (eV)	References
3. S=CPh₂ ... Ph, Ph + cyclopentadiene	(thietane, S, Ph, Ph)	7.6	(9, 10)
4. dione + MeHC=CHMe	cis: (Me, Me) 57% + (Me, Me) 43%; trans: (Me, Me) 57% + (Me, Me) 43%	8.2	(11)
		8.2	(11)
5. ketone (Me, Me)* + EtHC=CHOMe	cis: (Me Me, Et OMe) + (Me Me, Et OMe); trans: " + "	8.3	(12)
		8.3	(12)
6. aldehyde (Et)* + CH₂=CH–OEt	(oxetane, Et, OEt) Major	8.6	(13)

Table 20 (continued)

Reactants	Products	$I_D - A_A$ (eV)	References
7. 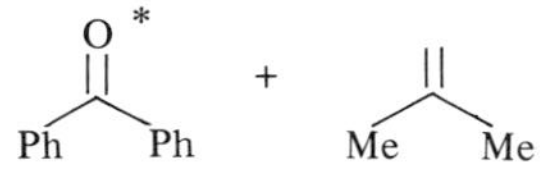	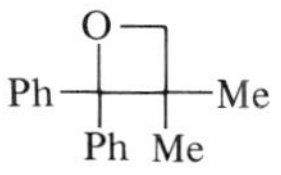	8.6	(14)

C. ππ* Singlet photocycloadditions

Reactants	Products	$I_D - A_A$ (eV)	References
1. + $MeO_2CCH=CHCO_2Me$	cis: 98% trans: 98%	6.6	(15)
2. + $MeO_2CCH=CHCO_2Me$	cis: 18% trans: 18% + 12% 12%	7.0 7.0	(16) (16)

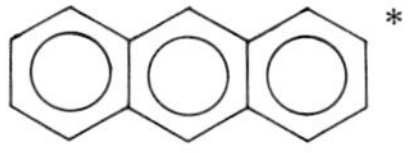

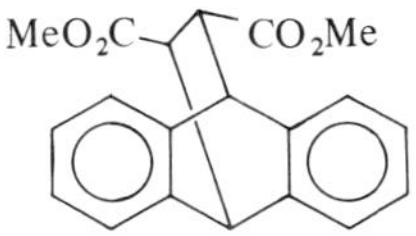

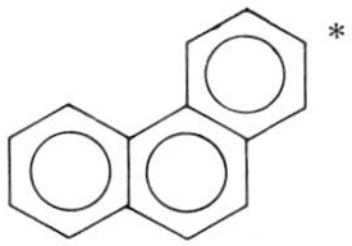

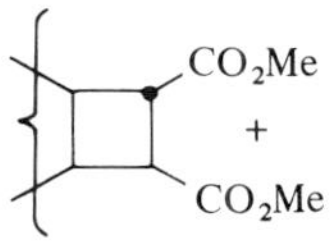

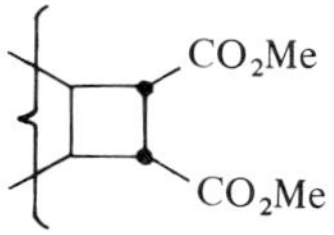

131

Table 20 (continued)

	Reactants	Products	$I_D - A_A$ (eV)	References
3.	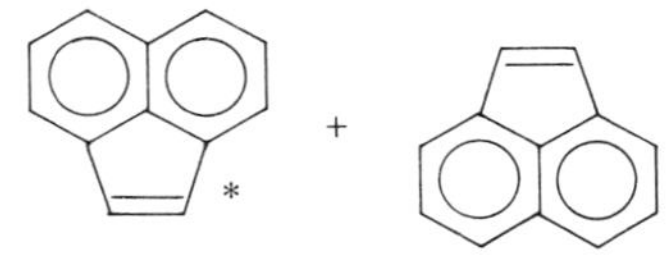	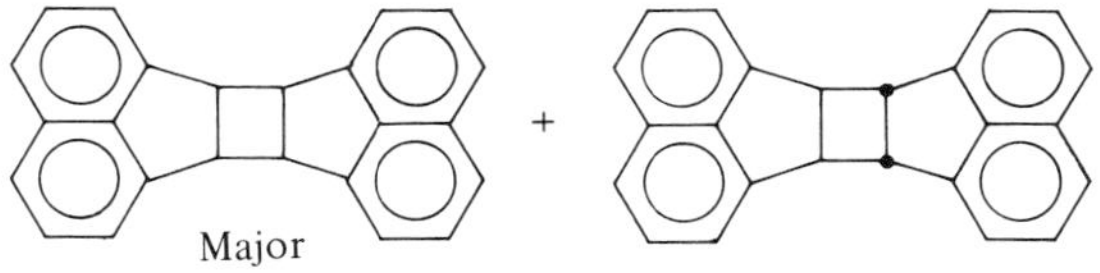 Major	7.4	(17)
4.	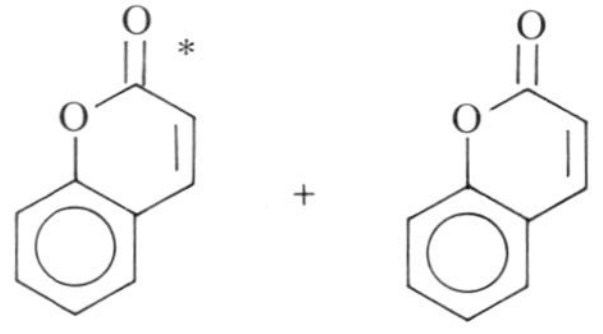	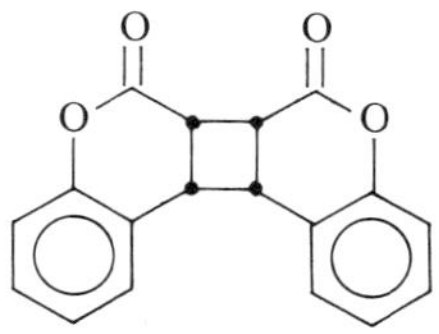	8.5	(18)
5.		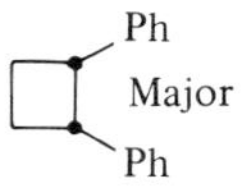 Major	8.7	(19)
6.		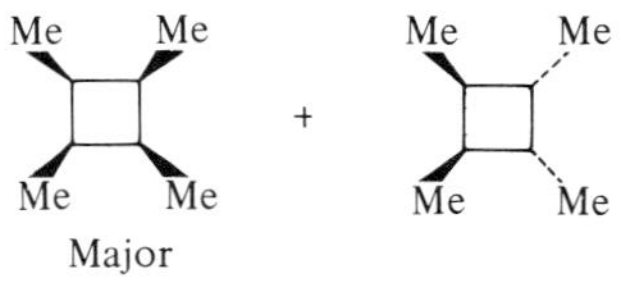 Major	11.0	(20)

132

Table 20 (continued)

Reactants	Products	$I_D - A_A$ (eV)	References
7. Me–CH=CH–Me + Me–CH=CH–Me	Me Me + Me Me (cyclobutane products)	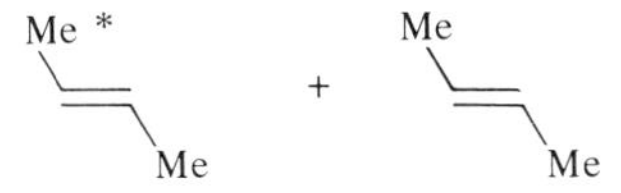 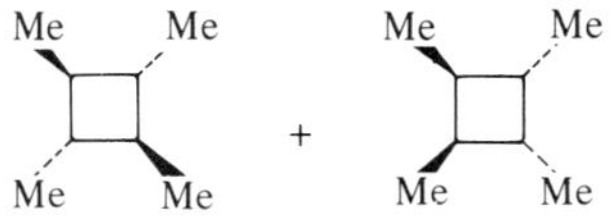 11.0	(20)
8. OPh* + OPh  	OPh OPh 17.1% + OPh OPh 12.9%	10.8	(21)

D. $\pi\pi^*$ Triplet photocycloadditions

1. phenanthrene* + MeO$_2$CHC=CHCO$_2$Me 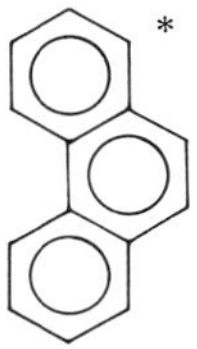 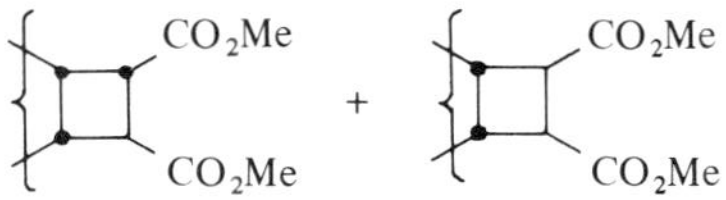	trans: CO$_2$Me CO$_2$Me 66% + CO$_2$Me CO$_2$Me 33%	7.0	(22)
2. indene* + CN	(CN cyclobutane product)	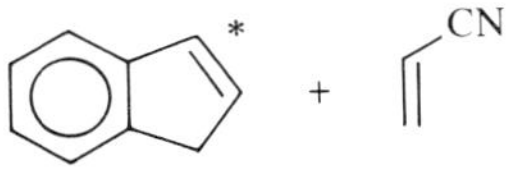 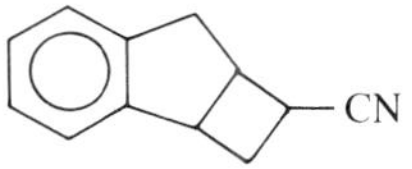 8.0	(23)

133

Table 20 (continued)

Reactants	Products	$I_D - A_A$ (eV)	References
3.	cis:	8.5	(24)
	+ trans:	8.5	(24)
4.	Major	8.7	(19)
5.	Major	8.8	(25)

Table 20 (continued)

Reactants	Products	$I_D - A_A$ (eV)	References
6. 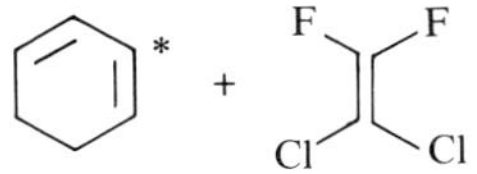 cyclohexene* + F₂C=CClCl	72% 3%	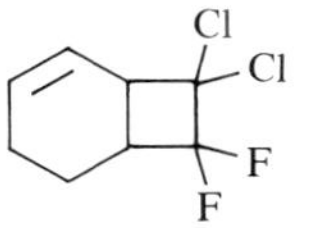 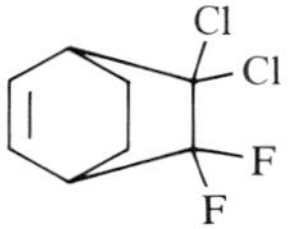 8.8	(26)
7. indene* + Cl HC=CCl H cis: −60 °C trans: −40 °C	cis: 73.4% 26.6% trans: 14.9% 85.1%	9.0 9.0	(27) (27)
8. 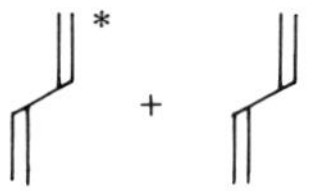 butadiene* + butadiene	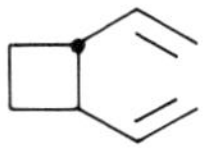 Major + ...	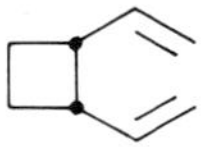 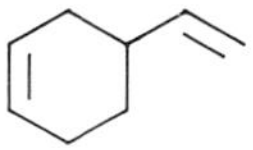 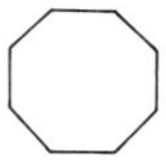 9.5	(28)

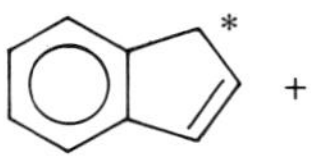

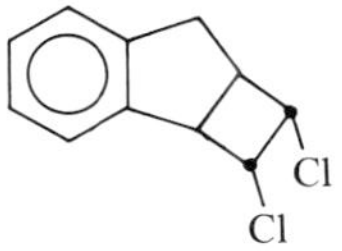

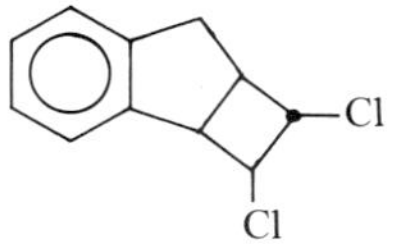

135

Table 20 (continued)

Reactants	Products	$I_D - A_A$ (eV)	References
9.		10.4	(29)

E. Photocycloadditions of unknown multiplicity

Reactants	Products	$I_D - A_A$ (eV)	References
1.	88% + 12%	5.4 $n\pi^*$	(30)
2. + $RO_2CCH=CHCO_2R$* (cis or trans)	Major	8.0 $\pi\pi^*$	(31)

Table 20 (continued)

Reactants	Products		$I_D - A_A$ (eV)	Multiplicity	References

F. Photocycloadditions of unknown state

1.

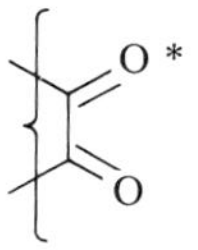

cis 15°C: 72% + 28%
70°C: 60% + 40%

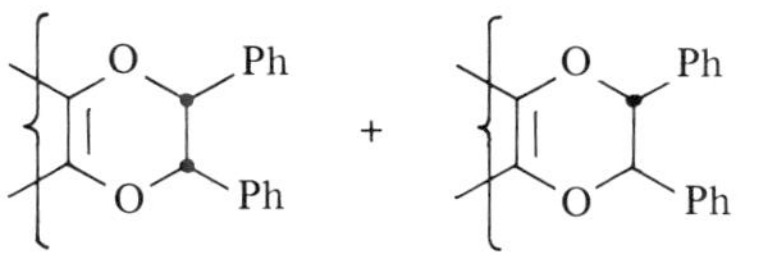

trans 15°C: 19% + 81%
70°C: 35% + 65%

6.8 — Triplet — (32)

2.

49% + 21%

6.8 — Triplet — (32)

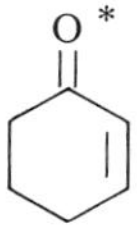

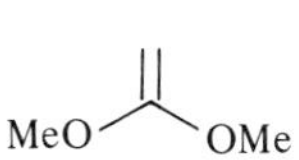

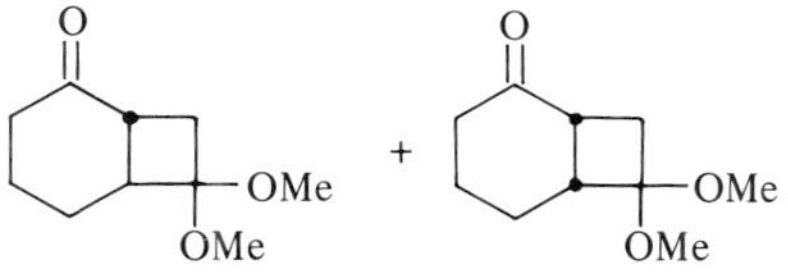

8.0 — Triplet — (33)

3.

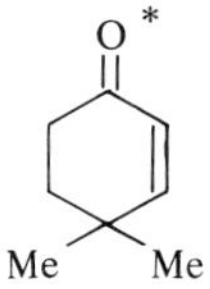

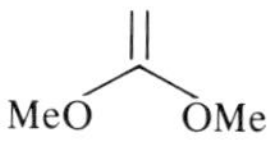

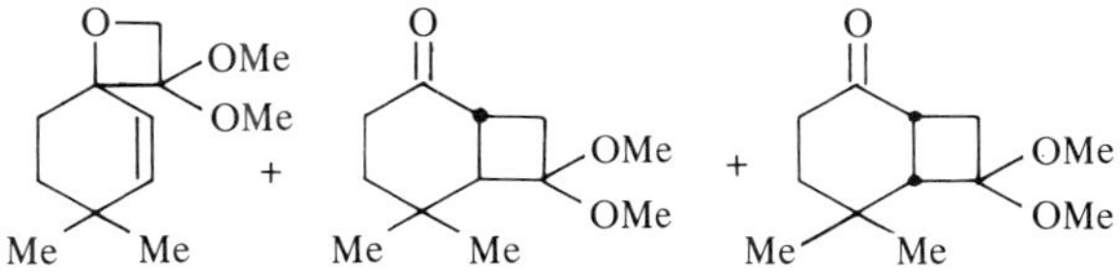

8.1 — Triplet — (34)

137

Reactivity Trends of Singlet Photochemical Cycloadditions

Table 20 (continued)

Reactants	Products	$I_D - A_A$ (eV)	Multiplicity	References
4.	42% + 47%	8.3	Triplet	(35)
5. R = CO₂Me	35% + 47% + 18%	8.5	Triplet	(36)

Table 20 (continued)

	Reactants	Products	$I_D - A_A$ (eV)	Multiplicity	References
6. 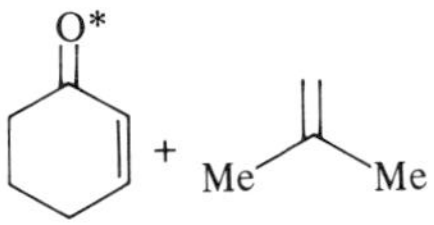		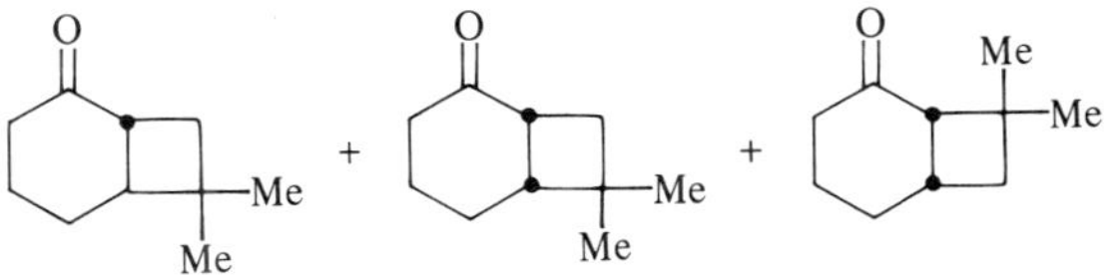 26.5% + 6.5% + 6%	9.2	Triplet	(33)
7. 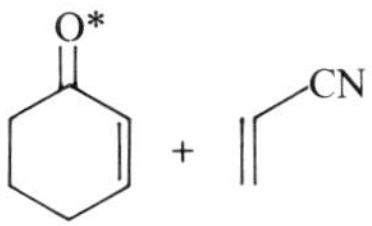			10.9	Triplet	(33)

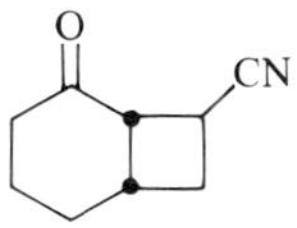

References see page 140

139

(1) Barltrop, J. A., Hesp, B.: J. Chem. Soc. *1965*, 5182.

(2) Lenz, G. R.: Tetrahedron *28*, 2211 (1972).

(3) Rettig, T. A.: Ph. D. Thesis, Iowa State Univ., Ames 1965.

(4) Hautala, R. R., Dawes, K., Turro, N. J.: Tetrahedron Let. *1972*, 1229.

(5) Barltrop, J. A., Carless, H. A. J.: J. Am. Chem. Soc. *94*, 1951 (1972).

(6) Dalton, J. C., Wriede, P. A., Turro, N. J.: J. Am. Chem. Soc. *92*, 1318 (1970).

(7) Scharf, H. D.: Tetrahedron Let. *1967*, 4231.

(8) Schenck, G. O., Kuhls, J., Krauch, C. H.: Ann. *693*, 20 (1966).

(9) Ohno, A.: Int. J. Sulfur Chem. B, *6*, 183 (1971).

(10) Yamada, K., Yoshioka, M., Sugiyama, N.: J. Org. Chem. *33*, 1240 (1968).

(11) Chow, Y. L., Joseph, T. C.: Chem. Commun. *1968*, 604.

(12) Turro, N. J., Wriede, P. A.: J. Am. Chem. Soc. *92*, 320 (1970).

(13) Shima, K., Sakurai, H.: Bull. Chem. Soc. Japan *42*, 849 (1969).

(14) Yang, N. C.: Pure Appl. Chem. *9*, 591 (1964).

(15) Kaupp, G.: Angew. Chem. Int. Edit. Engl. *11*, 313 (1972).

(16) Kaupp, G.: Angew. Chem. Int. Edit. Engl. *12*, 765 (1973).

(17) a) Cowan, D. O., Drisko, R. L.: Tetrahedron Let. *1967*, 1255.
b) Livingston, R., Wei, K. S.: J. Phys. Chem. *71*, 541 (1967).

(18) Hammond, G. S., Stout, C. A., Lamola, A. A.: J. Am. Chem. Soc. *86*, 3103 (1964).

(19) Brown, W. G.: J. Am. Chem. Soc. *90*, 1916 (1968).

(20) Yamazaki, H., Cvetanović, R. J., Irwin, R. S.: J. Am. Chem. Soc. *98*, 2198 (1976).

(21) Kuwata, S., Shigemitsu, Y., Odaira, Y.: Chem. Commun. *1972*, 2.

(22) a) Farid, S., Doty, J. C., Williams, J. L. R.: Chem. Commun. *1972*, 711.
b) Caldwell, R. A.: J. Am. Chem. Soc. *95*, 1690 (1973).

(23) Bowman, R. M., McCullough, J. J., Swenton, J. S.: Can. J. Chem. *47*, 4503 (1969).

(24) Schenck, G. O., Hartmann, W., Steinmetz, R.: Chem. Ber. *96*, 498 (1963).

(25) Schenck, G. O., Mannsfeld, S. P., Schomburg, G., Krauch, C. H.: Z. Naturforsch. *19b*, 18 (1964).

(26) Turro, N. J., Bartlett, P. D.: J. Org. Chem. *30*, 1849 (1965).

(27) Metzner, W.: Tetrahedron Let. *1968*, 1321.

(28) Hammond, G. S., Turro, N. J., Fischer, A.: J. Am. Chem. Soc. *83*, 4674 (1961); Liu, R. S. H., Turro, N. J., Hammond, G. S.: J. Am. Chem. Soc. *87*, 3406 (1965).

(29) Wagner, P. J., Bucheck, D. J.: J. Am. Chem. Soc. *91*, 5090 (1969).

(30) Bryce-Smith, D., Gilbert, A.: Chem. Commun. *1968*, 1701.

(31) Cox, A., de Mayo, P., Yip, R. W.: J. Am. Chem. Soc. *88*, 1043 (1966).

(32) Farid, S.: Chem. Commun. *1967*, 1268.

(33) Corey, E. J., Bass, J. D., LeMahieu, R., Mitra, R. B.: J. Am. Chem. Soc. *86*, 5570 (1964).

(34) Chapman, O. L., Koch, T. H., Klein, F., Nelson, P. J., Brown, E. L.: J. Am. Chem. Soc. *90*, 1657 (1968).

(35) Nelson, P. J., Ostrem, D., Lassila, J. D., Chapman, O. L.: J. Org. Chem. *34*, 811 (1969).

(36) Kaufmann, H. P., Gupta, A. K. S.: Ann. *681*, 39 (1965).

7.3 "Unusual" Head to Head vs. Head to Tail Regioselectivity of Photocycloadditions

The HH versus HT regioselectivity of certain photocycloadditions as predicted by our approach is opposite to predictions based on the "diradical" hypothesis. According to this latter hypothesis, photocycloadditions involve the intermediacy of "diradicals" and their regioselectivity is the one consistent with the formation of the most stable "diradical". Experimental results which contradict these intuitive expectations and are consistent with our predictions are cited in Table 20, e.g., entry A6.

7.4 nπ^* State Dual Channel Mechanisms of Polar Nonionic Carbonyl Photocycloadditions

The photocycloaddition of nπ^* excited enone to electron rich olefins may constitute a case where the concepts of Sect. 4.12 are applicable. Specifically, these photoreactions may involve an nπ^* state dual channel mechanism where the lowest energy path leads to inefficient $_2\pi_s + _2\pi_s$ cycloaddition and energy wastage via an excited intermediate while the higher energy path leads to $_2\pi_s + _2\pi_a$ cyclobutane and oxetane formation.

The following experimental evidence is consistent with the above expectations:

a) The photocycloaddition of enone type molecules to electron donor olefins leads to the predominant formation of the thermodynamically unfavorable trans fused adduct which can be viewed as the result of $_2\pi_s + _2\pi_a$ addition.

b) The formation of the trans fused adduct becomes increasingly favored as the donor ability of the olefin or the acceptor ability of the enone increases.

These trends are exemplified by reference to entries A2, A3, and A5 in Table 20.

In addition, a number of polar nonionic photocycloadditions involving an nπ^* excited carbonyl or dicarbonyl and a ground state diene or olefin are known to yield $4\pi + 2\pi$ adducts (e.g., entry A1 in Table 20). Such results signal an nπ^* state dual channel mechanism where the lowest energy path leads to energy wastage and inefficient $_2\pi_s + _2\pi_s$ cycloadduct formation via an excited intermediate and the higher energy path involving $_4\pi_s + _2\pi_s$ and $_2\pi_s + _2\pi_a$ cycloaddition is responsible for major product formation.

7.5 $\pi\pi^*$ State Dual Channel Mechanisms of Polar Nonionic Photocycloadditions

Due to a low ionization potential and a relatively substantial electron affinity, anthracene is predisposed to undergo highly polar nonionic cycloadditions with olefins and aromatic molecules. Accordingly, the photocycloadditions of $\pi\pi^*$ excited anthracene[1] and its derivatives may well involve dual channel mechanisms and the study of such

---

1 Naphthalene and anthracene have *two* close lying $\pi\pi^*$ singlets, one involving HO → LU (L_a) and the other subHO → LU and HO → subLU (L_b) promotions, Since L_a absorbs strongly and L_b absorbs weakly, we assume that, in experiments where the wavelength of the exciting light is not precisely defined, L_a is the photochemically active $\pi\pi^*$ singlet state. A similar assumption is made for *substituted* benzenes in Chap. 12.

reactions is of pivotal importance. A substantial body of experimental data has been amassed and the key trends are discussed below.

For the purpose of illustrating the key concepts, consider the reaction shown in Scheme 2 and assume that product **A** is consistent with minimization and product **B**

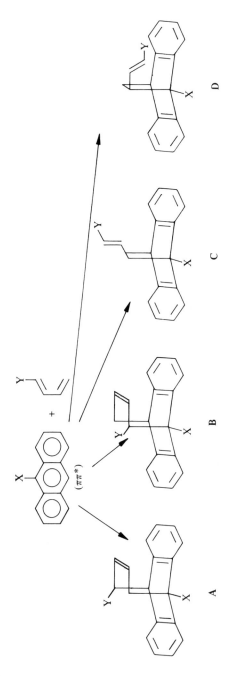

Scheme 2[a]

[a] The two additional possible $4\pi + 2\pi$ regioisomers have been omitted.

consistent with maximization of the $4\pi + 4\pi$ $HO^D - LU^A$ overlap integral between reactants while product **C** is consistent with minimization and product **D** with maximization of the $4\pi + 2\pi$ $HO^D - LU^A$ overlap integral between reactants. Furthermore, we shall assume that the various products are consistent with a reverse optimization of the $HO^D - HO^A$ and $LU^D - LU^A$ overlap integrals between reactants. On the basis of the theoretical principles outlined before, the following projections can be made:

a) Predominant formation of product **D** is consistent with significant energy wastage and a dual channel mechanism. The same will be true if the major reaction product is **C** and coulombic effects responsible for the alternate regiochemistry.

b) Predominant formation of product **A** is consistent with negligible energy wastage and a normal mechanism. The same will be true if the major reaction product is **B** and coulombic effects responsible for the alternate regiochemistry. In the absence of significant coulombic effects, formation of **B** may signal energy wastage and a dual channel mechanism.

c) As reaction polarity increases by appropriate choice of the X and Y substituents, a dual channel mechanism will become increasingly probable and major product formation will shift from **A** (and/or **B**) to **D** (and/or **C**).

In accordance with the above expectations, the following experimental results have been obtained:

a) In the photocycloaddition of anthracene derivatives and dienes, and increase in polarity leads to a change from $4\pi + 4\pi$ to $4\pi + 2\pi$ union (see Table 21).

b) In the photocycloaddition of anthracene derivatives and polyenes which proceed in a $4\pi + 2\pi$ manner, the preferred regioisomer is the one consistent with maximization of the $HO^D - LU^A$ overlap integral between reactants. Typical examples are entries 1 and 2 of Table 22.

Table 21. The regiochemistry of polar nonionic photocycloadditions of polynuclear hydrocarbons[a]

CHD[b]	100% $4\pi + 4\pi$	100% $4\pi + 4\pi$	100% $4\pi + 4\pi$	67% $4\pi + 4\pi$
				9% $4\pi + 2\pi$
DMHD[b]	100% $4\pi + 4\pi$	100% $4\pi + 4\pi$	100% $4\pi + 2\pi$	100% $4\pi + 2\pi$

[a] Yang, N. C., Srinivasachar, K., Kim, B., Libman, J.: J. Amer. Chem. Soc. 97, 5006 (1975).
[b] CHD = Cyclohexadiene, DMHD = 2,4-dimethylhexadiene.

144

Table 22. Photoadditions of anthracene derivatives

Reactants	Chorochemistry[a]	Overlap integral				References
		$HO^D - LU^A$	$HO^A - LU^D$	$HO^D - HO^A$	$LU^D - LU^A$	
1. CN anthracene + $(CH_3)_2C=CH-CH=C(CH_3)_2$	$4\pi_s + 4\pi_s$	0.0654	0.0020	0.4716	0.5406	(1)
	$4\pi_s + 2\pi_s$ HH (boxed)	0.4499	0.3811	0.0881	0.1561	
	$4\pi_s + 2\pi_s$ HT	0.4249	0.3821	0.0919	0.0506	
2. CN anthracene + cycloheptatriene	$4\pi_s + 4\pi_s$ HH	0.0048	0.0424	0.3693	0.4174	(2)
	$4\pi_s + 4\pi_s$ HT	0.0976	0.0386	0.3688	0.4286	
	$4\pi_s + 2\pi_s$ HH	0.3234	0.2965	0.1151	0.0891	
	$4\pi_s + 2\pi_s$ HT (boxed)	0.3549	0.2954	0.1121	0.1712	
	$6\pi_s + 4\pi_s$	0.4694	0.0021	0.0021	0.0631	

a The major product is indicated in rectangle. In all cases, the 9 and 10 positions of anthracene are involved in the cycloaddition.

(1) Yang, N. C., Libman, J., Barrett Jr., L., Hui, M. H., Loeschen, R. L.: J. Amer. Chem. Soc. 94, 1406 (1972).
(2) Yang, N. C., Srinivasachar, K.: Chem. Commun. 1976, 48.

An important extension of the concept of $\pi\pi^*$ state-dual channel polar photo-cycloadditions is illustrated by considering the reaction of **D** and **A***. As we have discussed, the path which minimizes the interaction of DA* (and D*A) and D$^+$A$^-$ diabatic surfaces and maximizes that of D$^+$A$^-$ and DA will become favored over the alternative path which has opposite consequences. When minimization of the HO^D − HO^A and LU^D − LU^A interactions and maximization of the HO^D − LU^A are not simultaneously possible in a $_2\pi_s$ + $_2\pi_a$ or $_4\pi_s$ + $_2\pi_s$ approach, the next best choice is incomplete optimization via a TB stereochemical approach leading to nonstereoselective product formation. This could explain the nonstereoselectivity of certain polar singlet photocycloadditions (see also Sect. 7.7 and 21.5).

Finally, before we conclude the discussion of the experimental results regarding photocycloadditions, it should be pointed out that several types of excited intermediates (i.e. M, N*, etc.) which are predicted to be involved in such reactions have been detected or inferred experimentally and the list continues to grow [56]. Furthermore, the analysis of photocycloadditions presented in Chap. 4 leads to an additional interesting prediction regarding the structure of excited intermediates involved in photoreactions. Specifically, it is expected that the intermolecular distance of a stable excited intermediate (**D**...**A**)* will be longer than that between **D** and **A** at the transition state of the corresponding thermal reaction. The observation that typical "excimers" involve only a small degree of overlap [57] of the two partners seems to indicate that our expectations are not unreasonable[1].

In closing, the following cautionary remarks are in order:

a) Ionic photocycloadditions have not been discussed simply because there is not available experimental data with which to compare our predictions. The reasons are obvious: thermal ionic reactions are very fast and cannot be easily suppressed so that the photochemical reaction is investigated without complications due to competing thermal processes.

b) The HH versus HT regioselectivity of photochemical cycloadditions may be dictated by coulombic effects if the "rate determining step" of the reaction, i.e., photochemical barrier crossing or decay, occurs at very long intermolecular distances where "orbital effects" are deemphasized and coulombic effects attending HH and HT union are very significantly different.

c) As we have discussed before, photocycloadditions may involve the formation of TB intermediates which can undergo bond rotation leading to limited or extensive stereorandomization. However, nonstereoselectivity may also be the result of competing pathways involving multicentric orbital overlap. In the space below, I illustrate how such a situation may arise.

Consider the polar photoreactions shown below involving an excited acceptor molecule and a ground state donor partner and proceeding via a dual channel mecha-

---

1   It should be emphasized that the stability of the excited intermediate of a photochemical reaction predicted by our approach reflects mainly the enthalpic factor. Entropy effects may augment or reduce the stability of this intermediate.

nism. Neglecting steric effects, both reactions are expected to occur in a $_2\pi_s + {_2}\pi_a$ manner, where the acceptor plays the role of the antarafacial component. However, steric effects dictate that such a stereochemical union is favorable in the case of the cis

geometric isomer and unfavorable in the case of the trans one. Hence, the photocyclo-addition of the cis isomer may lead to predominant formation of 2 via a $_2\pi_s + {_2}\pi_a$ process while the trans isomer may lead also to formation of 2 via the less efficient $_2\pi_s + {_2}\pi_s$ process.

In other words, here we have a situation where, starting from two different geometric isomers, we can obtain the same product mixture. Currently, this is taken to signify the intermediacy of a freely rotating molecular species. However, competing pathways subject to different steric constraints may also account for such a result.

146

8. Miscellaneous Intermolecular Multicentric Reactions

8.1 Cycloadditions of Cumulene Systems

A special $2\pi + 2\pi$ cycloaddition involves one component of the type $R_2C=C=X$. The prototype reaction given below illustrates our approach. The MO's of $CH_2=C=X$, where $X=O$, are shown in Fig. 41.

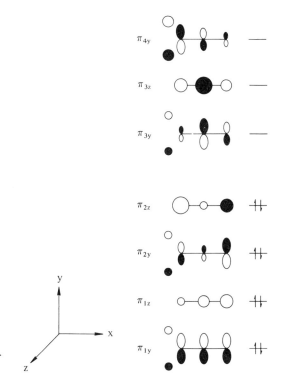

Fig. 41. Pi type MO's of ketene computed by the CNDO/2 method

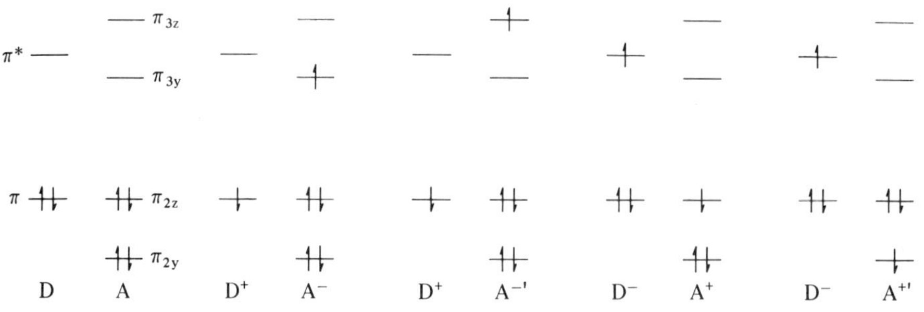

π^* with π_{3z}, π_{3y} levels and π with π_{2z}, π_{2y} levels shown for the configurations:

	D	A	D⁺	A⁻	D⁺	A⁻'	D⁻	A⁺	D⁻	A⁺'

Fig. 42. Zero order configurations for the description of $2\pi + 2\pi$ cycloaddition of an acceptor cumulene and a donor olefin

The five configurations which one has to consider in discussing the reactivity aspects of the thermal $2\pi + 2\pi$ reaction which proceeds via an aromatic path are shown in Fig. 42. In a reaction where $CH_2=C=X$ acts as the acceptor, the two lowest charge transfer configurations are D^+A^- and $D^+A^{-'}$. Since the $DA-D^+A^{-'}$ interaction enforces $_2\pi_s + _2\pi_a$ pericyclic bonding while the $DA-D^+A^-$ dominant interaction enforces three center bonding involving the two vinylic carbons of the olefin and the central carbon of $H_2C=C=X$, the pericyclicity of the ground adiabatic surface will depend upon the relative strength of the two interactions. We distinguish the following cases:

a) X is such so that the $\pi_{3y} - \pi_{3z}$ energy gap is small, in which case the reaction complex will resemble a true $_2\pi_s + _2\pi_a$ complex.

b) X is such so that the $\pi_{3y} - \pi_{3z}$ energy gap is large, in which case the reaction will resemble a $_2\pi_s + _2\pi_a$-like intermediate.

c) For a fixed X, increasing donor strength of the olefin will tend to accentuate the dominant interaction and render the electronic characteristics of the reaction complex similar to those expected from a $_2\pi_s + _2\pi_a$-like intermediate.

In $2\pi + 2\pi$ cycloaddition where $CH_2=C=X$ acts as a donor, the $LU^A - \pi_{2z}$ interaction becomes dominant; the second most important interaction is $LU^A - \pi_{2y}$. By contrast to the previous case, the dominant interaction enforces pericyclicity and the alternative interaction predominant three center bonding.

An interesting corollary of the above analysis is that an acceptor $CH_2=C=X$ molecule will react with an acceptor diene in a preferred $_2\pi_s + _2\pi_a$ manner, because a $_4\pi_s + _2\pi_s$ geometry renders the dominant $HO^D - \pi_{3y}$ interaction zero. On the other hand, a donor $CH_2=C=X$ molecule can react with an acceptor diene in a $_4\pi_s + _2\pi_s$ manner because the dominant $LU^A - \pi_{2z}$ interaction is maximized in this geometry.

The preferred regiochemistry of the $_2\pi_s + _2\pi_a$ cycloaddition where $CH_2=C=X$ acts as an acceptor will be the one which involves union of the central carbon of $CH_2=C=X$ with the vinylic carbon of the donor olefin having highest HO electron density. On the other hand, the preferred regiochemistry of the $_2\pi_s + _2\pi_a$ or $_4\pi_s + _2\pi_s$

cycloaddition where $CH_2=C=X$ acts as a donor will be the one which maximizes $\pi_{2z} - LU^A$ overlap.

Finally, the rates of $CH_2=C=X$ $_2\pi_s + _2\pi_a$ cycloaddition will be determined conjointly by steric effects and reaction polarity.

Experimental results concerning ketene cycloadditions are in good harmony with the above predictions:

a) $2\pi + 2\pi$ cycloadditions where $RR'C=C=O$ acts as the acceptor are s + a stereoselective [58]. If ketene plays the role of the acceptor, $2\pi + 2\pi$ dominates $4\pi + 2\pi$ addition [59].

b) Cycloaddition where $RR'C=C=O$ acts as the donor and the diene as the acceptor yield $4\pi + 2\pi$ adducts. Examples are shown below [60, 61].

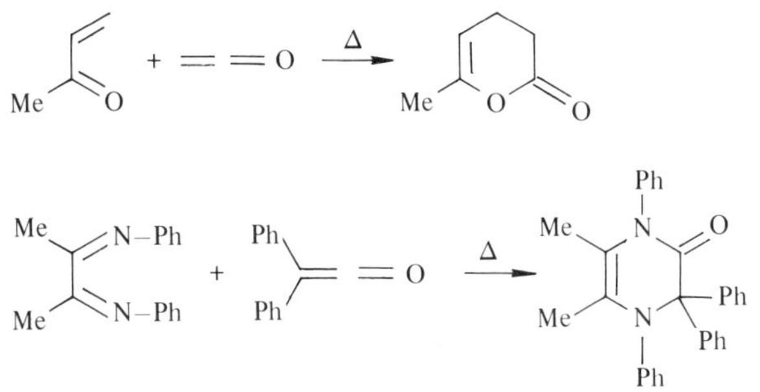

c) The regioselectivity of thermal $_2\pi_s + _2\pi_a$ cycloadditions where $RR'C=C=O$ acts as an acceptor is as predicted. An example is given below [59].

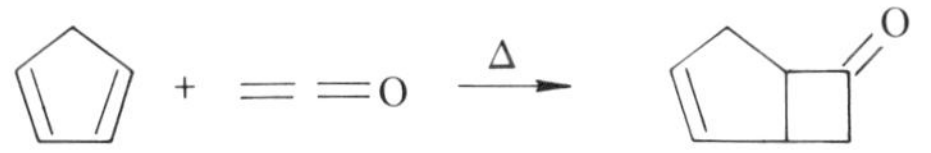

Similarly, the regioselectivity of thermal $_4\pi_s + _2\pi_s$ cycloadditions where $RR'C=C=O$ acts as a donor is as expected [e.g., the reactions under (b)].

Endo product formation in the cycloaddition of an unsymmetrically substituted ketene and a donor olefin demonstrates the steric requirement of the $_2\pi_s + _2\pi_a$ cyclo-

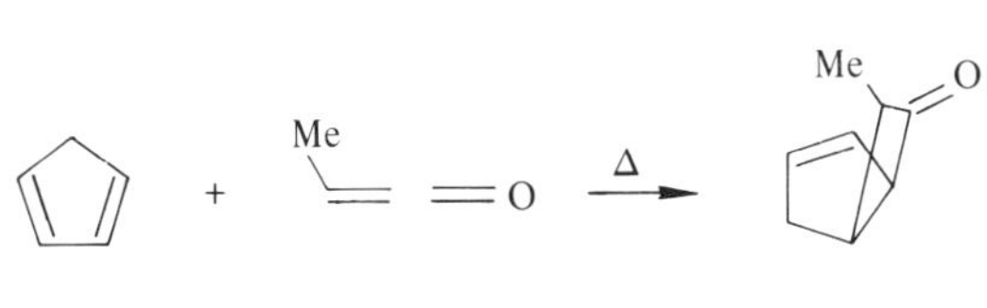

addition [62–65]. An example is given below [66]. Endo regioselectivity is due to the minimization of steric repulsion as is well-known.

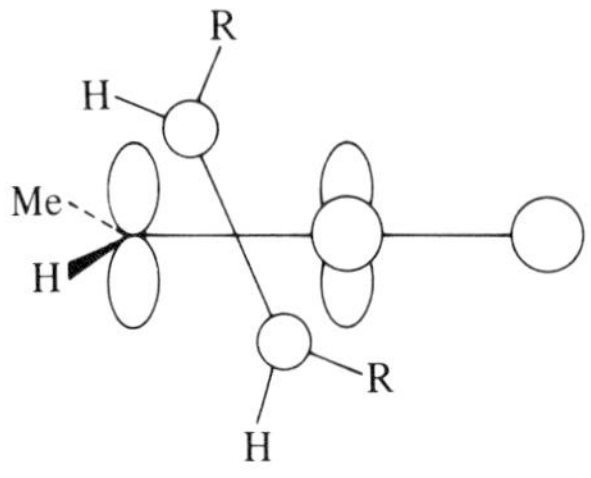

d) The rates of $_2\pi_s + _2\pi_a$ cycloaddition of $R'RC=C=O$ are strongly affected by the steric requirement of the olefin partner. An example is [67]:

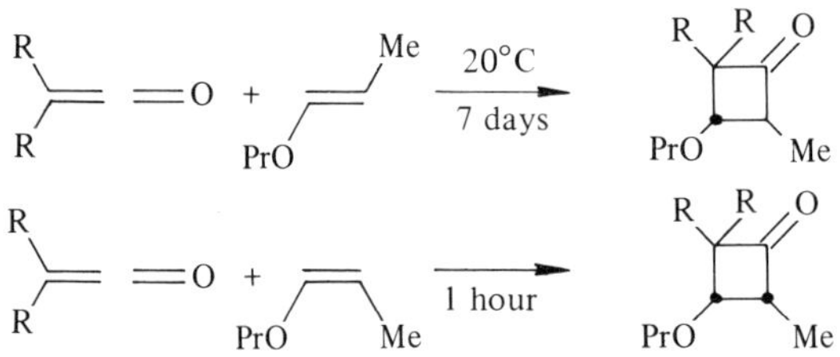

The above discussion has been applicable to thermal nonionic cycloadditions. Thermal ionic cycloadditions can be analyzed similarly. In such reactions, an N* intermediate will form which could be trapped. Indeed, this is what experimental results suggest [68]. The N* intermediate will have a $_2\pi_s + _2\pi_a$ geometry.

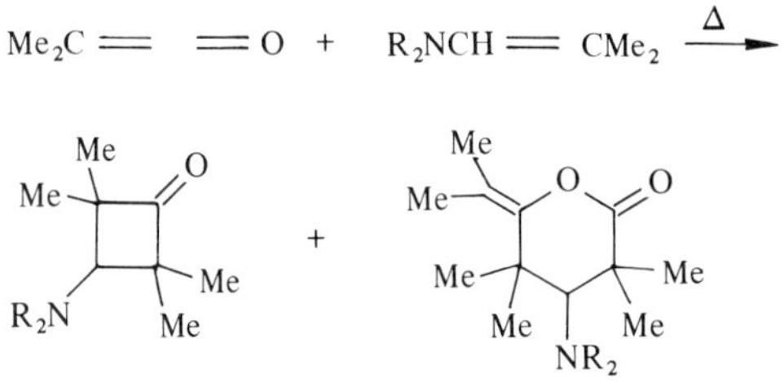

$$Me_2C =\!\!=\!\!= O \ + \ R_2NCH =\!\!= CMe_2 \ \xrightarrow{\Delta}$$

8.2 The Ene Reaction

The ene reaction can be exemplified by the chemical equation shown below. The pi type MO's of propene are "similar" to those of 1,3-butadiene as illustrated in Fig. 43.

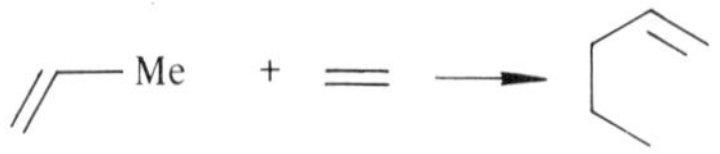

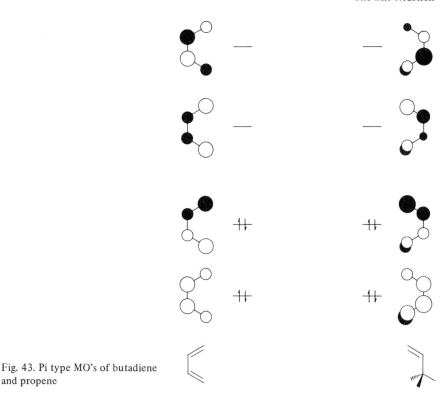

Fig. 43. Pi type MO's of butadiene and propene

Accordingly, the analysis of $4\pi + 2\pi$ cycloadditions are also applicable to ene reactions, as the following experimental results show:

a) Thermal ene reactions are particularly facile when an electron acceptor olefin acts as the reaction partner. This is due to the intrinsic donor character of the ene component.

b) The regioselectivity of thermal ene reactions is, in general, the one consistent with maximization of the $HO^D - LU^A$ interaction. An example is given below [69].

$$CH_3CH = CH_2 \ + \ CH_2 = CHCO_2Me \ \xrightarrow{\Delta} \ $$

88%   12%

c) Low polarity ene photoreactions do not occur, i.e. a $2\pi + 2\pi$ or, in general, a 4N electron photocycloaddition predominates. However, an increase in polarity leads to the realization of the ene photoreaction. In the examples given below [70, 71], it is

151

very likely that triplet excitation is responsible for reaction. The intriguing shift from a 4N to a 4N + 2 electron triplet photoreaction which accompanies an increase in polarity is discussed in Chap. 19.

8.3 1,3 Dipolar Cycloadditions

1,3 dipolar cycloadditions are $4\pi + 2\pi$ cycloadditions. In general, one may distinguish

between 1,3 dipoles of the allyl anion type ($X=\overset{+}{Y}-\bar{Z}$ and $\overset{+}{X}-Y-\bar{Z}$) and those of the propargenyl anion type ($X\equiv\overset{+}{Y}-\bar{Z}$ and $\overset{+}{X}=Y-\bar{Z}$), Calculations show that 1,3 dipoles like $>C=\underset{|}{C}-\bar{C}<$ and $-C\equiv\overset{+}{N}-\bar{C}<$ have high lying HO's and are predisposed to act as donors while 1,3 dipoles like $O=\overset{+}{O}-\bar{O}$ have low lying LU's and are predisposed to act as acceptors [42]. This transition is easy to understand since replacement of carbon by a more electronegative first row atom tends to depress the energies of occupied as well as unoccupied MO's.

The effect of polarity on the rates of thermal 1,3 dipolar cycloadditions is in accordance with our predictions. Specifically, when a 1,3 dipole predisposed to act as a donor is employed in the reaction (e.g., $PH_2\bar{C}-\overset{+}{N}\equiv N$), the rate increases as the acceptor ability of the dipolarophile increases [72]. On the other hand, when a 1,3 dipole predisposed to act as an acceptor participates in the reaction (e.g., O_3) the rate increases as the donor ability of the dipolarophile increases [73].

The choroselectivity of 1,3 dipolar cycloadditions (thermal) are consistent with expectations based on our analysis [49].

152

Photochemical 1,3 dipolar cycloadditions have not been studied systematically. However, 1,3 dipolar photocycloadditions occur when reaction polarity is high. The examples given below [74–76] involve triplet excited states. The all-important effect of reaction polarity on triplet photoreactivity is discussed in Chap. 19.

Proposed intermediate

9. $\pi + \sigma$ Addition Reactions

9.1 Introduction

A typical $\pi + \sigma$ addition reaction is shown below. In general, $\pi + \sigma$ additions can be thought of as belonging to the general class of cycloaddition reactions and, accord-

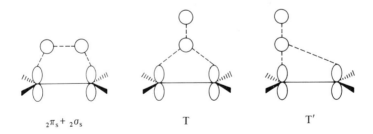

ingly, can be treated in a similar fashion. The most interesting aspects of these reactions are the following:

a) *Stereoselectivity.* We need to consider only three distinct stereochemical modes of reaction, namely, $_2\pi_s + {_2}\sigma_s$, T, and T$'$. The alternative $_2\pi_s + {_2}\sigma_a$ and $_2\pi_a + {_2}\sigma_s$ reaction modes are very unfavorable due to poor spatial orbital overlap. Note that

$$_2\pi_s + {_2}\sigma_s \qquad T \qquad T'$$

the $_2\pi_s + {_2}\sigma_s$ geometry leads directly to products while the T and T$'$ geometries do not. Hence, the $_2\pi_s + {_2}\sigma_s$ stereochemical mode is comparable to the $_2\pi_s + {_2}\pi_s$ stereochemical mode and the T and T$'$ modes comparable to the TB mode in $2\pi + 2\pi$ cycloadditions.

b) *Regioselectivity.* When the unsaturated system and the addend are both unsymmetrically substituted, regiochemical preference for an addition mode will arise.

c) Effect of reaction polarity on reaction rates.

154

9.2 Potential Energy Surfaces for $2\pi + 2\sigma$ Additions

There are three types of $2\pi + 2\sigma$ addition reactions which one needs to consider. The first is exemplified by the reaction of ethylene and hydrogen molecule. In this case the two pi electrons of the ethylene and the two sigma electrons of hydrogen molecule are the ones which are crucial for the stereochemical outcome of the reaction. The P.E. surfaces for $2\pi_s + 2\sigma_s$ addition are similar to those for $2\pi_s + 2\pi_s$ cycloaddition and the P.E. surfaces for T or T' addition are similar to those for TB cycloaddition.

The ultimate stereochemical consequences of the $2\pi_s + 2\sigma_s$, T and T' paths can be easily understood in the case of thermal $2\pi + 2\sigma$ addition. By following the same type of reasoning as before, we conclude that the $2\pi_s + 2\sigma_s$ path will proceed with distortion and lead to cis addition while the T and T' paths will produce an intermediate which will be attacked by yet another molecule of H_2 to yield a final trans addition product.

The second type of $2\pi + 2\sigma$ addition is exemplified by the reaction of ethylene and hydrogen fluoride. In this case, the two pi electrons of ethylene, the two sigma bonding electrons of hydrogen fluoride and one lone pair of fluorine are all involved in the reaction. Thus, this reaction provides an example of the role of lone pairs. The $2\pi_s + 2\sigma_s$, T and T' geometries are shown below. The reshuffling of sigma bond

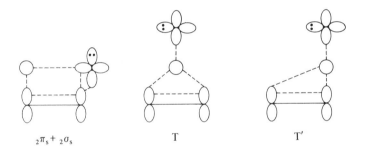

$2\pi_s + 2\sigma_s$ T T'

and lone pair electrons is a key feature of the reaction. It follows that the basis set configurations should include the no bond, monoexcited and diexcited configurations involving the two pi electrons of $CH_2=CH_2$, the two sigma electrons of HF *and* the appropriate lone pair electrons of F. The configurations constituting the Λ_1 and Λ_2 packets are shown in Fig. 44. The reader can easily construct the configurations which comprise the Λ_3 packet. The intrapacket interactions are of the $HO^D - HO^A$, $LU^D - LU^A$, and $n-HO^D$ type while the interpacket interactions are of the $HO^D - LU^A$, $HO^A - LU^D$ and $n-LU^D$ type. Because of the asymmetry of HF, both inter- and intrapacket interactions will play a role in determining the surface features for $2\pi_s + 2\sigma_s$ addition. Of course, the same is true for T and T' addition. The incursion of the lone pair does not alter the qualitative shapes of the P.E. surfaces which will be comparable to those for the previously discussed case. It merely introduces new interaction matrix elements which can come into play in determining the regioselectivity of certain $2\pi + 2\sigma$ additions.

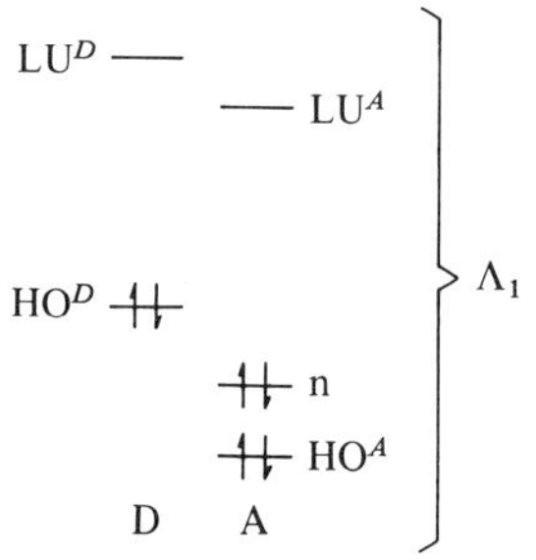

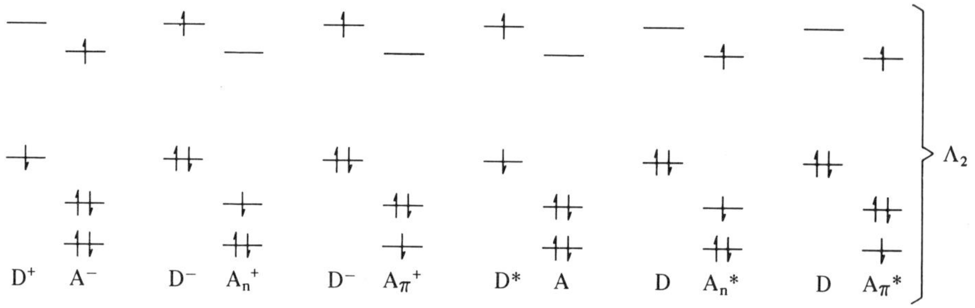

Fig. 44. Typical zero order configurations for the treatment of a $2\pi + 2\sigma$ addition involving a lone pair of one reactant. Λ_1 and Λ_2 are packets

As in the case of $\pi + \pi$ cycloaddition, the key features of the $_2\pi_s + _2\sigma_s$ addition can be understood by restricting attention to the DA, D^+A^-, and D^*A diabatic surfaces plus the boundary of the Λ_3 packet, which is made up primarily of the D^*A^* diabatic surface. Similarly, the key features of the T or T′ addition can be understood by restricting attention to the DA, D^+A^-, D^*A, and D^*A^* diabatic surfaces. The most interesting stereochemical prediction is that as reaction polarity increases the point will be reached when the reaction becomes ionic. In this event, the $_2\pi_s + _2\sigma_s$ path will not be separated by a large energy gap from the T or T′ path. In fact, since the $_2\pi_s + _2\sigma_s$ mechanism leads directly to product formation, it may very well become the preferred mechanism.

The third type of $2\pi + 2\sigma$ additions reactions is exemplified by the reaction of ethylene and molecular fluorine. In this case, two pi, two sigma and four lone pair electrons are involved in the reaction and the basis set configurations must be constructed accordingly. By going through familiar arguments, we can predict that the lone pairs will not alter the qualitative aspects of P.E. surfaces expected for a thermal antiaromatic ($_2\pi_s + _2\sigma_s$) and a nonaromatic (T or T′) pathway. Once again, the most interesting prediction is that increased polarity will eventually lead to effectively pericyclic $_2\pi_s + _2\sigma_s$ reaction.

156

From the above discussion, it becomes apparent that the regioselection rules for $2\pi + 2\sigma$ additions will be the same as those for $2\pi + 2\pi$ cycloadditions with the additional feature of lone pair-HO and lone pair-LU interactions becoming important in appropriately designed systems. Finally, the reaction toposelectivity and the effect of polarity on reaction rates can be discussed in the usual manner.

9.3 Reactivity Trends of $2\pi + 2\sigma$ Additions

Most $2\pi + 2\sigma$ addition reactions occurring in solution involve initial electrophilic or nucleophilic attack of an unsaturated system by a solvent polarized or ion pair-like addend. In general $2\pi + 2\sigma$ additions in solution are ionic reactions. A T or T′ mechanism is expected to lead to an ultimate trans addition while a ${}_2\pi_s + {}_2\sigma_s$ mechanism to cis addition. Thus, it is expected that ionic $2\pi + 2\sigma$ additions will be trans. However, as the degree of polarity increases a turnover to a cis mechanism will occur. Experimental support of these ideas does exist. Thus, while additions of HCl, HBr, and Cl_2 to simple olefins occur in a preferred trans manner, additions of the same molecules to arenes and good electron donor olefins occur in a preferred cis manner. Representative examples are shown below.

$\xrightarrow[\text{AcOH}]{\text{DBr}}$ 76% trans + 24% cis [77]

$\xrightarrow[\text{CH}_2\text{Cl}_2]{\text{DBr}}$ 15% trans + 85% cis [78]

$\xrightarrow{\text{Cl}_2}$ 100% trans [79]

$\xrightarrow[\text{CCl}_4]{\text{Cl}_2}$ 40% trans + 60% cis [80]

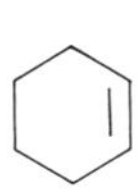

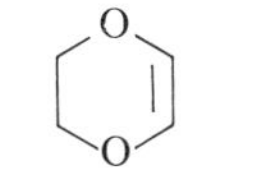

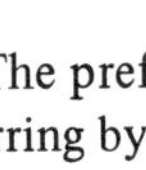

The preferred regiochemistry of ionic $2\pi + 2\sigma$ additions in solution, whether occurring by any of the ${}_2\pi_s + {}_2\sigma_s$, T or T′ mechanisms, is predicted to be the one which maximizes the $HO^D - LU^A$ interaction.

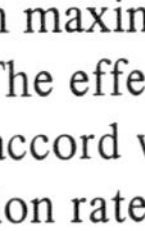

The effect of reaction polarity on the rates of ionic $2\pi + 2\sigma$ additions in solution is in accord with our predictions, i.e., a significant increase in polarity enhances the reaction rate. Correlations of this type can be found in various review articles [81]. The interesting orbital electron density "dilution" effect can be discerned in cases where polarity differences are small. Typical examples are given below [81c].

$$PhCH=CH_2 + Br_2 \rightarrow PhCHBrCH_2Br \qquad k_{rel} = 1.0$$

$$MeCH=CHMe + Br_2 \rightarrow MeCHBrCHBrMe \qquad k_{rel} = 19.5$$

In gas phase $2\pi + 2\sigma$ retroadditions, a significant polarity increase leads to a rate enhancement and the regioselectivity of the reaction appears to be consistent with maximization of the $HO^D - LU^A$ interaction. An example is given below [82].

$$Me-\underset{\underset{Cl}{|}}{\overset{\overset{Me}{|}}{C}}-Me \quad \xrightarrow[\text{kcal/mole}]{E_a = 45.0} \quad (Me)_2C = CH_2 + HCl$$

$$Me-\underset{\underset{H}{|}}{\overset{\overset{Me}{|}}{C}}-CH_2Cl \quad \xrightarrow[\text{kcal/mole}]{E_a = 56.9} \quad (Me)_2C = CH_2 + HCl$$

Another general type of reaction which can formally be classified as a nonionic $2\pi + 2\sigma$ addition is the one shown below. The strained sigma bond constitutes the

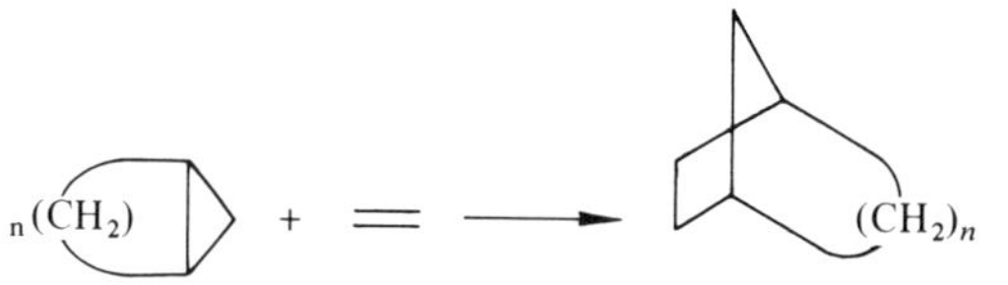

donor partner in its cycloadditions with olefins of varying electronic nature. As the electrophilicity of the olefin partner increases, cis addition should become increasingly prominent and the rate of the reaction should also increase. Indeed, it was found that increasing electrophilicity of the olefin partner led to enhanced reaction rates [83, 84]. The following order of reactivity was established for various substituted acetylenes and ethylenes:

$$NCC{\equiv}CCN \quad > \quad MeOOCC{\equiv}CCOOMe \quad > \quad PhC{\equiv}CPh$$

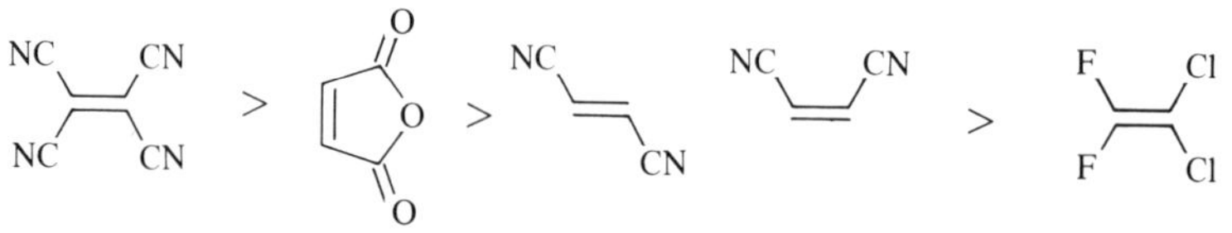

It was also found that the reaction of bicyclo[2.1.0]pentane with either fumaronitrile or maleonitrile is highly stereoselective.

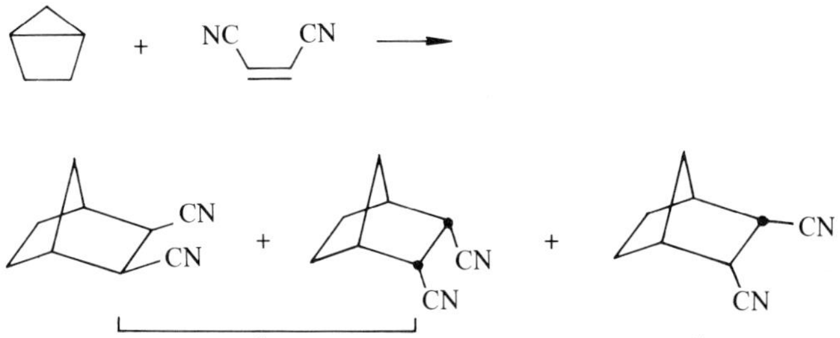

5 : 1

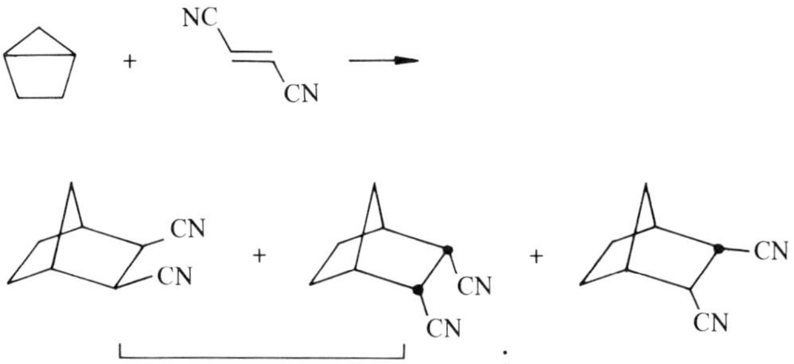

2.3 : 22

10. Even-Odd Multicentric Intermolecular Reactions

10.1 Potential Energy Surfaces for $2\pi + 1\pi$ Cycloadditions

An interesting area of chemical research where there has not been much theoretical guidance is the study of the reactions of radical ions which can be formed chemically or generated in a spectrometer, e.g., the mass spectrometer. In this chapter, we consider the stereoselectivity and toposelectivity aspects of such reactions.

We shall illustrate our approach by examining the model reaction shown below.

$$= \ + \ \overset{\cdot}{=} \ \longrightarrow \ \square^{\overset{\cdot}{+}}$$

The zero order configurations necessary for the description of this reaction are shown in Fig. 45. Proper symmetry adaptation leads to the following equations:

$$\Psi_1^+ = \frac{1}{\sqrt{2}} \ (D_1 D_2^+ + D_1^+ D_2) \tag{10.1}$$

$$\Psi_1^- = \frac{1}{\sqrt{2}} \ (D_1 D_2^+ - D_1^+ D_2) \tag{10.2}$$

$$\Psi_2^+ = \frac{1}{\sqrt{2}} \ (D_1^* D_2^+ + D_1^+ D_2^*) \tag{10.3}$$

$$\Psi_2^- = \frac{1}{\sqrt{2}} \ (D_1^* D_2^+ - D_1^+ D_2^*) \tag{10.4}$$

$$\Psi_3^+ = \frac{1}{\sqrt{2}} \ (D_1^- D_2^{++} + D_1^{++} D_2^-) \tag{10.5}$$

$$\Psi_3^- = \frac{1}{\sqrt{2}} \ (D_1^- D_2^{++} - D_1^{++} D_2^-) \tag{10.6}$$

$$\Psi_4^+ = \frac{1}{\sqrt{2}} \ (D_1 D_2^{+*} + D_1^{+*} D_2) \tag{10.7}$$

$$\Psi_4^- = \frac{1}{\sqrt{2}} \ (D_1 D_2^{+*} - D_1^{+*} D_2) \tag{10.8}$$

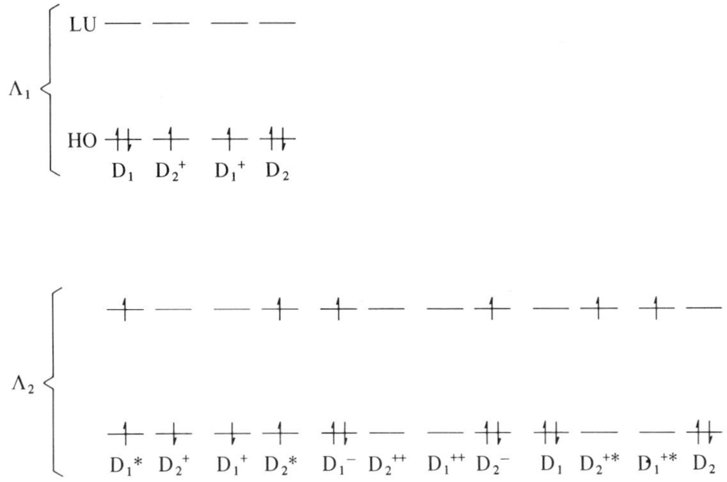

Fig. 45. Zero order configurations for the treatment of the $2\pi + 2\pi$ cycloaddition of ethylene and ethylene cation radical

The equations for the diabatic surfaces are:

$$\epsilon(\Psi_1^+) = \epsilon(\Psi_1^-) \simeq S \tag{10.9}$$

$$\epsilon(\Psi_2^+) = \epsilon(\Psi_2^-) = \epsilon(D_1^* D_2^+) \simeq {}^3G(HO \to LU) + S' \tag{10.10}$$

$$\epsilon(\Psi_3^+) = \epsilon(\Psi_3^-) = \epsilon(D_1^- D_2^{+2}) \simeq I^C - A + C + S' \tag{10.11}$$

$$\epsilon(\Psi_4^+) = \epsilon(\Psi_4^-) = \epsilon(D_1 D_2^{+*}) \simeq {}^2G(HO \to LU) + S' \tag{10.12}$$

The interaction matrix is given in Table 23.

Due to symetry adaptation, the energies of the diagonal elements Ψ_1^+, Ψ_1^-, Ψ_2^+ and Ψ_2^- depend upon orbital overlap via $HO^{D_1} - HO^{D_2}$ and $LU^{D_1} - LU^{D_2}$ type interaction matrix elements. The intrapacket interactions of Λ_1 are zero while those of Λ_2 are of the $HO^{D_1} - HO^{D_2}$ and $LU^{D_1} - LU^{D_2}$ type. Finally, the interpacket interactions are of the $HO^{D_1} - LU^{D_2}$ and $HO^{D_2} - LU^{D_1}$ type. By following familiar arguments, we conclude that in the case of $_2\pi_s + _1\pi_s$ cycloaddition there are strong diagonal and intrapacket interactions and zero interpacket interactions, while in the case of a $_2\pi_s + _1\pi_a$ cycloaddition there are strong interpacket interactions and zero diagonal and intrapacket interactions.

The diabatic surfaces are shown in Fig. 46. In the case of $_2\pi_s + _1\pi_s$ cycloaddition, the boundary of the Λ_2 packet will overtake the boundary of the Λ_1 packet. However, unlike the case of the $2\pi + 2\pi$ cycloaddition of two closed shell ethylenes, the boundaries of both Λ_1 and Λ_2 move towards lower energy as the intermolecular distance of the two reaction systems decreases. The adiabatic surfaces are shown in

Table 23. Interaction matrix for the cycloaddition of ethylene and ethylene radical cation[a]

	Ψ_1^+	Ψ_1^-	Ψ_2^+	Ψ_2^-	Ψ_3^+	Ψ_3^-	Ψ_4^+	Ψ_4^-
Ψ_1^+	$-(HO^{D_1}-HO^{D_2})$	0	$HO^{D_1}-LU^{D_2}+$ $HO^{D_2}-LU^{D_1}$	0	$LU^{D_1}-HO^{D_2}+$ $LU^{D_2}-HO^{D_1}$	0	0	0
Ψ_1^-		$+(HO^{D_1}-HO^{D_2})$	0	$HO^{D_1}-LU^{D_2}+$ $HO^{D_2}-LU^{D_1}$	0	$LU^{D_1}-HO^{D_2}+$ $LU^{D_2}-HO^{D_1}$	0	0
Ψ_2^+			$-(LU^{D_1}-LU^{D_2})$	0	$HO^{D_1}-HO^{D_2}$	0	$HO^{D_1}-HO^{D_2}$	0
Ψ_2^-				$+(LU^{D_1}-LU^{D_2})$	0	$HO^{D_1}-HO^{D_2}$	0	$HO^{D_1}-HO^{D_2}$
Ψ_3^+					—	0	$LU^{D_1}-LU^{D_2}$	0
Ψ_3^-						—	0	$LU^{D_1}-LU^{D_2}$
Ψ_4^+							—	0
Ψ_4^-								—

a   There exist two doublet states for three open shell electrons. Only one of the two possible wavefunctions is considered in this work.

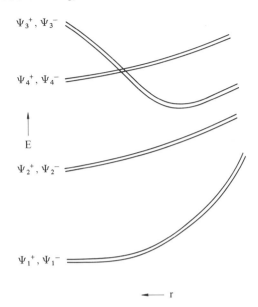

Ψ_3^+, Ψ_3^-

Ψ_4^+, Ψ_4^-

E

Ψ_2^+, Ψ_2^-

Ψ_1^+, Ψ_1^-

Fig. 46. Diabatic surfaces for the treatment of the $2\pi + 2\pi$ cycloaddition of ethylene and ethylene cation radical

← r

Fig. 47. In the case of $_2\pi_s + _1\pi_a$ cycloaddition, the interpacket interaction will determine the final shapes of the adiabatic surfaces which are shown in Fig. 47.

From the discussion above, it is apparent that a satisfactory understanding of any $_2\pi_s + _1\pi_s$ cycloaddition can be achieved by considering only the no-bond configurations of Λ_1 and the boundary of the Λ_2 packet. Furthermore, a satisfactory understanding of any $_2\pi_s + _1\pi_a$ (or $_2\pi_a + _1\pi_s$) cycloaddition can be achieved by considering the no bond configurations of Λ_1 as well as the locally excited configurations of Λ_2 which can interact with those of Λ_1 (see Fig. 48).

On the basis of the P.E. surfaces of Fig. 48, the following conclusions regarding thermal $2\pi + 1\pi$ cycloadditions can be reached:

a) According to the TRR, the $_2\pi_s + _1\pi_a$ (or $_2\pi_a + _1\pi_s$) path will be favored electronically over the $_2\pi_s + _1\pi_s$ path. However, the difference of the barriers of these two paths is expected to be small due to the presence of intrapacket interactions (diagonal interactions) in Λ_1 in the latter case. As a result, steric effects may well render the $_2\pi_s + _1\pi_s$ path the preferred reaction path. Furthermore, an EE $_2\pi_s + _2\pi_s$ process will be more unfavorable than the corresponding EO $_2\pi_s + _1\pi_s$ process[1].

---

1 This is due to the interaction of the D^+A-DA^+ diabatic surfaces in EO reactions. The interaction of the $DA-D^+A^-$ in the corresponding EE reactions is zero or small.

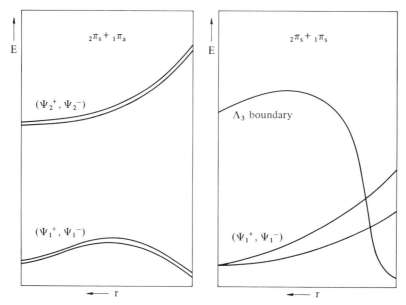

Fig. 47. Potential energy surfaces for $_2\pi_s + _1\pi_a$ and $_2\pi_s + _1\pi_s$ cycloaddition of ethylene and ethylene cation radical. The pertinent diabatic surfaces at infinite intermolecular distance are indicated in parentheses

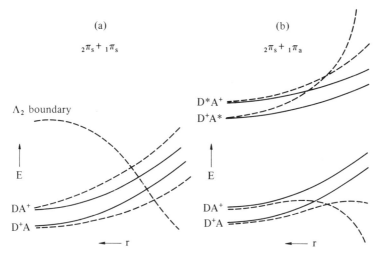

Fig. 48. Diabatic (solid lines) and adiabatic (dashed lines) surfaces for $_2\pi_s + _1\pi_s$ and $_2\pi_s + _1\pi_a$ cycloaddition

b) The $_2\pi_s + _1\pi_s$ process will occur regiochemically in such a manner as to maximize intrapacket interactions, which are of the HO—HO type, while the $_2\pi_s + _1\pi_a$ (or $_2\pi_a + _1\pi_s$) process will occur in such a manner as to maximize interpacket interactions, which are of the HO—LU type.

The various reactivity aspects of photochemical $2\pi + 1\pi$ cycloadditions can be discussed also in the usual manner. The following trends are noted:

a) A photochemical $2\pi + 1\pi$ cycloaddition initiated via excitation of the neutral olefin will occur in a preferred $_2\pi_s + _1\pi_s$ manner because this stereochemistry allows for a low photochemical barrier and efficient decay to the ground state product.

b) The regiochemistry of a photochemical $_2\pi_s + _1\pi_s$ cycloaddition will be such as to maximize intrapacket interactions and minimize interpacket interactions.

The discussion provided above is appropriate to reactions of a radical cation generated by ejection of an electron in vacuum. In solution, the radical cation is generated in the vicinity of a counterion. Accordingly, in this event, the basis set configurations should reflect the presence of the counterion. This can be achieved by replacing D^+A by X^-D^+A, DA^+ by X^-DA^+, etc., assuming that the counterion X^- remains in the vicinity of D^+. Under these circumstances, the P.E. surfaces derived assuming absence of the counterion will be somewhat modified.

Once again, the $_2\pi_s + _1\pi_s$ and $_2\pi_s + _1\pi_a$ ($_2\pi_a + _1\pi_s$) thermal cycloadditions are nothing else but models for pseudoantiaromatic and pseudoaromatic thermal paths, respectively.

The ideas presented in this section can be extended to any EO $\pi + \pi$ cycloaddition in the manner indicated below.

No. of Pi electrons	Stereochemical path	P.E. surface
4N + 1	s + a	Fig. 48b
4N + 1	s + s	Fig. 48a
4N − 1	s + a	Fig. 48b
4N − 1	s + s	Fig. 48a

Extension to EO $\pi + \sigma$ additions is trivial.

10.2 Reactivity Trends of Cationic Even-Odd Retro-Cycloadditions and Eliminations

Experimental support for the idea that multicentric EO reactions occurring in the mass spectrometer are thermal pseudoantiaromatic processes, which are much more favorable than the corresponding multicentric EE reactions, can be found in a recent study of formal $2\pi + 1\pi$ (or $1\pi + 2\pi$) cycloreversions. In this work, the 1,2 elimination of molecular hydrogen from a radical cation was considered in the light of Woodward-Hoffmann correlation diagrams and the suggestion was made that "forbidden" 1,2 elimination of hydrogen will occur with the release of a substantial amount of kinetic energy [85]. In odd electron systems, however, our approach suggests that a $_2\pi_s + _1\pi_s$ or $_1\pi_s + _2\pi_s$ retroaddition is modestly favorable and, in any event, more favorable than the corresponding EE reaction. From Table 24, the dra-

165

matic increase in the kinetic energy release when the reacting ion changes from an odd electron system to an even electron system is readily apparent. In the former case, loss of H_2 is an s + s retroaddition of a neutral molecule and a cation radical which is a modestly favorable reaction according to our theory. By contrast, in the latter case, loss of H_2 is an s + s retroaddition of two closed shell molecules which is an unfavorable reaction. The experimental activation energies for the eliminations shown in Table 25 confirm this clear cut pattern which is only explicable if H_2 elimination from a radical ion occurs in an s + s manner.

As pointed out by Williams and Hvistendahl [86], these activation energies are insufficient to produce either the neutral hydrogen or ionic molecule in an excited state, implying that multicentric reactions in the mass spectrometer occur on the ground surface.

Table 24. Kinetic energy release in 1,2 elimination of hydrogen

Reacting ion	Product ion	Kinetic energy release (kcal/mol)	References
$CH_3CH_3^{+\cdot}$	$CH_2CH_2^{+\cdot}$	4.4	(1)
$H_2C=\overset{+}{O}H$	$HC\equiv O^+$	33	(2)
$H_2C=\overset{+}{N}H_2$	$HC\equiv\overset{+}{N}H$	20	(1)
$H_2C=\overset{+}{S}H$	$HC\equiv\overset{+}{S}H$	20	(1)

(1) Williams, D. H., Hvistendahl, G.: J. Amer. Chem. Soc. 96, 6753 (1974).
(2) Beynon, J. H., Fontaine, A. E., Lester, G. R.: Int. J. Mass Spectrom. Ion Phys. 1, 1 (1968).

Table 25. Activation energies for hydrogen elimination of cations[a]

	E_a (kcal/mol)
$H_3CCH_3^{+\cdot} \rightarrow CH_2CH_2^{+\cdot} + H_2$	13
$H_2C=\overset{+}{O}H \rightarrow HC=\overset{+}{O} + H_2$	80
$H_2C=\overset{+}{N}H_2 \rightarrow HC\equiv\overset{+}{N}H + H_2$	97

[a] See (1) in Table 24.

11. Potential Energy Surfaces for Odd-Odd Multicentric Intermolecular Reactions

In this section we turn our attention to OO multicentric reactions. Our approach will be exemplified by reference to the model reaction shown below:

We shall again be concerned with two distinct stereochemical paths which this reaction may follow, i.e., the $_1\pi_s + _1\pi_s$ and the $_1\pi_s + _1\pi_a$ paths.

The zero order configurations necessary for the description of this reaction are shown in Fig. 49. Proper symmetry adaptation leads to the configuration wavefunctions shown below.

$$\Psi_1 = D_1^+ D_2^+ \tag{11.1}$$

$$\Psi_2^+ = \frac{1}{\sqrt{2}} (D_1 D_2^{+2} + D_1^{+2} D_2) \tag{11.2}$$

$$\Psi_2^- = \frac{1}{\sqrt{2}} (D_1 D_2^{+2} - D_1^{+2} D_2) \tag{11.3}$$

$$\Psi_3^+ = \frac{1}{\sqrt{2}} (D_1^+ D_2^{+*} + D_1^{+*} D_2^+) \tag{11.4}$$

$$\Psi_3^- = \frac{1}{\sqrt{2}} (D_1^+ D_2^{+*} - D_1^{+*} D_2^+) \tag{11.5}$$

$$\Psi_4^+ = \frac{1}{\sqrt{2}} (D_1^* D_2^{+2} + D_1^{+2} D_2^*) \tag{11.6}$$

$$\Psi_4^- = \frac{1}{\sqrt{2}} (D_1^* D_2^{+2} - D_1^{+2} D_2^*) \tag{11.7}$$

167

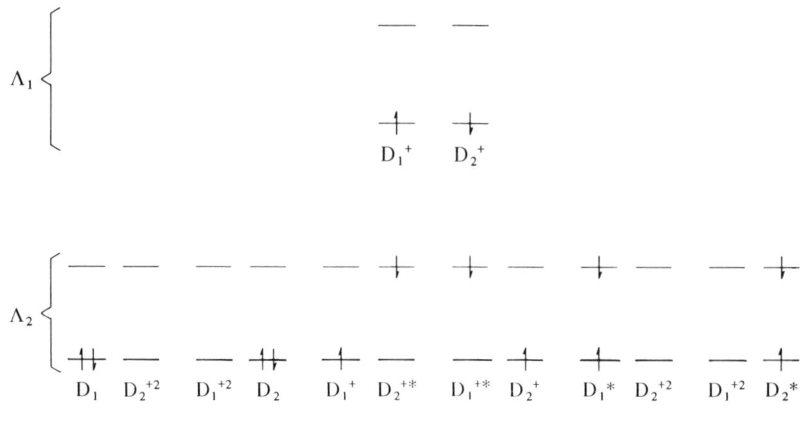

Λ_1 : D_1^+ D_2^+

Λ_2 : $D_1 \, D_2^{+2}$ $\quad$ $D_1^{+2} \, D_2$ $\quad$ $D_1^+ \, D_2^{+*}$ $\quad$ $D_1^{+*} \, D_2^+$ $\quad$ $D_1^* \, D_2^{+2}$ $\quad$ $D_1^{+2} \, D_2^*$

Fig. 49. Zero order configurations for $1\pi + 1\pi$ cyclodimerization of ethylene cation radical. Λ_1 and Λ_2 are packets

The interaction matrix is given in Table 26. It is clear that intrapacket interactions are of the HO–LU type and interpacket interactions of the HO–HO type. By following familiar arguments, we conclude that in the case of the $_1\pi_s + _1\pi_s$ cycloadditions there are strong interpacket interactions and zero intrapacket interactions, while in the case of a $_1\pi_s + _1\pi_a$ cycloaddition there are strong intrapacket interactions and zero interpacket interactions. In general, the $_1\pi_s + _1\pi_s$ cycloaddition can be simply described by considering the Ψ_1, Ψ_2^+, and Ψ_2^- diabatic surfaces, while the $_1\pi_s + _1\pi_a$ cycloaddition can be described by considering the Ψ_1 diabatic surface and the boundary of the Λ_2 packet.

At this point, the reader should note that the same pattern of diabatic surface interactions obtains in $_1\pi_s + _1\pi_s$ and $_2\pi_s + _2\pi_a$ cycloadditions, the former being a Hückel aromatic and the latter a Möbius aromatic reaction. A related analogy can be drawn between $_1\pi_s + _1\pi_a$ and $_2\pi_s + _2\pi_s$ cycloadditions, the former being a Möbius antiaromatic and the latter a Hückel antiaromatic reaction. Thus, it is obvious that the reaction of two cation radicals can be discussed in a way which is analogous to the one encountered in the case of the reaction of two even systems.

The combination of two cation or two anion radicals is energetically very unfavorable and will not occur under normal circumstances. On the other hand, anion radical $(A^{\pm})$-cation radical $(D^{+\cdot})$ combination reactions are feasible and, indeed, highly intriguing. These two molecular species can be produced either by ordinary chemical means or by electrochemical methods. The $D^{+\cdot} - A^{\pm}$ complex corresponds to a high energy state of the EE reaction of D and A. Accordingly, we can immediately write down the equations for the chemical reactions of $D^{+\cdot}$ and $A^{\pm}$ by reference to the appropriate P.E. surfaces.

As an example, let us consider the case of a nonionic reaction where the D^+A^- diabatic surface is crossed only by the DA^* diabatic surface. In such a case, the equations for the reaction of the $D^{+\cdot}$ and $A^{\pm}$ are as follows:

Table 26. Interaction matrix for the $2\pi + 2\pi$ cycloaddimerization of ethylene radical cation

	Ψ_1	Ψ_2^+	Ψ_2^-	Ψ_3^+	Ψ_3^-	Ψ_4^+	Ψ_4^-
Ψ_1	—	$HO^{D_1}-HO^{D_2}$	0	0	0	$HO^{D_1}-LU^{D_2}$ $+HO^{D_2}-LU^{D_1}$	0
Ψ_2^+		—	0	$HO^{D_1}-LU^{D_2}$ $+HO^{D_2}-LU^{D_1}$	0	0	0
Ψ_2^-			—	0	$HO^{D_1}-LU^{D_2}$ $+HO^{D_2}-LU^{D_1}$	0	0
Ψ_3^+				—	0	$LU^{D_1}-LU^{D_2}$ $+HO^{D_1}-HO^{D_2}$	0
Ψ_3^-					—	0	$LU^{D_1}-LU^{D_2}$ $+HO^{D_1}-HO^{D_2}$
Ψ_4^+						—	0
Ψ_4^-							—

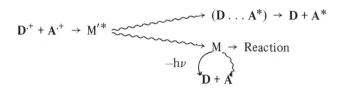

The equations above predict that radical ion pair annihilation leads to the formation of a photoexcited molecule, in addition to other possible products. This indeed has been observed experimentally [87].

12. Even-Even Intermolecular Bicentric Reactions

12.1 Potential Energy Surfaces for Electrophilic and Nucleophilic Additions

Ionic bicentric reactions constitute a most important class of organic reactions. Typical cases involve initial addition of a nucleophile (N^-) or electrophile (E^+) to an unsaturated system. The resulting adduct may then eliminate a chemical group giving rise to product formation. Typical examples are given below.

(a) $\quad$ ⟩=⟨ $\quad + N^-X^+ \quad \longrightarrow \quad$ ⟩—⟨ $\quad N \quad X^{\oplus} \quad \longrightarrow \quad$ Products

(b) $\quad$ ⟩=⟨ $\quad + E^+X^- \quad \longrightarrow \quad$ ⟩—⟨ $\quad E \quad X^{\ominus} \quad \longrightarrow \quad$ Products

$\quad$ Such reactions can be adequately understood by using a minimal basis set which includes the no bond configuration, the lowest energy charge transfer configuration and the locally excited configuration, where the excited partner is the unsaturated substrate. The basis sets for nucleophilic and electrophilic unions are constructed as follows:

a) Electrophilic Attack: DA, D^+A^- and D^*A where **D** is the unsaturated substrate and **A** the electrophilic ion pair E^+X^-.

b) Nucleophilic Attack: DA, D^+A^- and DA^* where **D** is the nucleophilic ion pair N^-X^+ and **A** the unsaturated substrate. The basis set configurations are shown in Fig. 50. Construction of the interaction matrices is trivial. In bicentric reactions, all one electron interaction matrix elements are large unless a nodal plane through a uniting center mandates otherwise. The equations of the diabatic surfaces are given below:

a) Electrophilic Attack

$$\epsilon(DA) \simeq S \tag{12.1}$$

$$\epsilon(D^+A^-) \simeq I_D - A_A + S' + C \qquad (12.2)$$

$$\epsilon(D^*A) \simeq {}^1G(HO^D \to LU^D) + S' \qquad (12.3)$$

b) Nucleophilic Attack

$$\epsilon(DA) \simeq S \qquad (12.4)$$

$$\epsilon(D^+A^-) \simeq I_D - A_A + S' + C \qquad (12.5)$$

$$\epsilon(DA^*) \simeq {}^1G(HO^A \to LU^A) + S' \qquad (12.6)$$

The P.E. surfaces for electrophilic and nucleophilic additions are shown in Fig. 51. The reaction mechanisms can be conveyed also by means of equations as follows:

Thermal: $\quad D + A \to (D \dots A) \to N_\pi^* \to N_\sigma^* \to$ Products

Photochemical:
$$
\begin{array}{l}
D^* + A \quad (D^* \dots A) \\
\text{or} \quad\quad \to \text{or} \quad\quad \to N^{**} \rightsquigarrow N_\pi^* \to N_\sigma^* \to \text{Products} \\
D + A^* \quad (D \dots A^*) \quad\quad\quad \searrow D + A
\end{array}
$$

In the case of type A reactions, formation of N_π^* is rate determining while in type B reactions, formation of N_σ^* is rate determining. By employing familiar arguments, we can predict the regioselectivity, polarity effect on reaction rate, and the selectivity-polarity relationship. The predictions are similar to the ones reached for type A and type B ionic aromatic multicentric reactions and are summarized in Table 27.

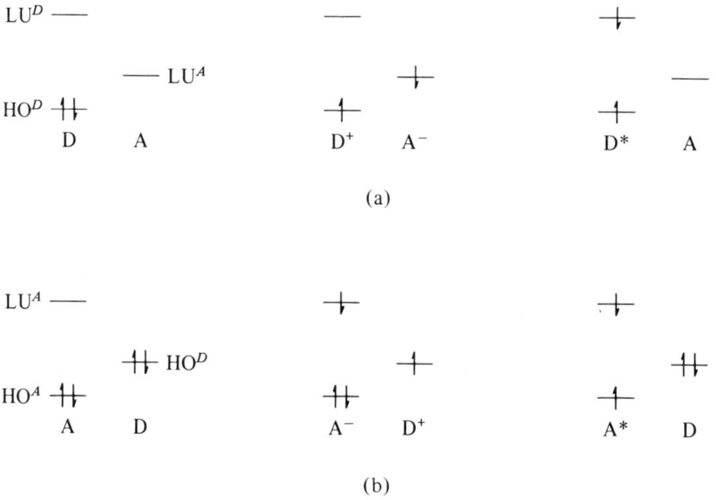

(a)

(b)

Fig. 50. Zero order configurations for a) electrophilic and b) nucleophilic additions to unsaturated systems

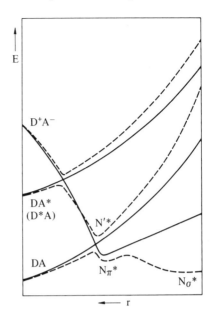

Fig. 51. Diabatic (solid lines) and adiabatic (dashed lines) surfaces for electrophilic or nucleophilic addition to an unsaturated system

Table 27. Reactivity aspects of electrophilic and nucleophilic even-even ionic additions

Model Reaction	Reaction Type	Regiochemical Condition			Selectivity[a] Response
		Ground Surface	Excited Surface	Decay Process	
Electrophilic Addition	Thermal, Type A	$HO^D\text{–}LU^A$	–	–	↓
	Thermal, Type B	$HO^D\text{–}LU^A$	–	–	↓
	Photochemical, Type A	–	$HO^D\text{–}HO^A$	$HO^D\text{–}LU^A(-)$	↓[b]
Nucleophilic Addition	Thermal, Type A	$HO^D\text{–}LU^A$	–	–	↓
	Thermal, Type B	$HO^D\text{–}LU^A$	–	–	↓
	Photochemical, Type A	–	$LU^D\text{–}LU^A$	$HO^D\text{–}LU^A(-)$	↓[b]

[a] An arrow pointing downwards denotes a decrease in selectivity as polarity increases, and conversely.

[b] Prediction valid under the assumption that the translation of a diabatic surface is not counteracted by the energy adjustment due to interaction.

Of particular importance are the regioselection rules for thermal and type A photochemical aromatic substitutions which can be stated as follows:

a) A thermal electrophilic aromatic substitution will proceed in such a manner as to maximize the $HO^D - LU^A$ matrix element. Hence, the regioselectivity of the reaction will be determined by the HO electron density of the aromatic (Aromatic = **D**).

b) A thermal nucleophilic aromatic substitution will also proceed in such a manner as to maximize the $HO^D - LU^A$ matrix element. Hence, the regioselectivity of the reaction will be determined by the LU electron density of the aromatic (Aromatic = **A**).

c) A photoelectrophilic aromatic substitution will proceed in such a manner as to maximize the $LU^D - LU^A$ matrix element and minimize the $HO^D - LU^A$ matrix element. Hence, the regioselectivity of the reaction will be determined jointly by the HO and LU electron density of the aromatic (Aromatic = **D**).

d) A photonucleophilic aromatic substitution will proceed in such a manner as to maximize the $HO^D - HO^A$ matrix element and minimize the $HO^D - LU^A$ matrix element. Hence, the regioselectivity of the reaction will be determined jointly by the HO and LU electron density of the aromatic (Aromatic = **A**).

12.2 Reactivity Trends of Electrophilic and Nucleophilic Substitutions

The P.E. surfaces for electrophilic and nucleophilic additions can provide the basis for discussing electrophilic and nucleophilic substitution reactions *insofar as the addition step is rate determining.* A survey of experimental results is presented below and is based upon the recognition of the following important facts:

a) The regioselectivity of a thermal substitution reaction is subject to the same controlling influence regardless of whether the reaction belongs to type A or type B.

b) A type A photosubstitution will exhibit the regioselectivity predicted in Table 27. On the other hand, a type B photosubstitution will need thermal activation for the conversion of the N_π^* to the N_σ^* complex and its regioselectivity may become identical to that of a thermal substitution.

c) The effect of polarity on reaction rates as well as the polarity-selectivity relationship is qualitatively the same for type A and type B thermal substitution reactions. Accordingly one could not differentiate type A from type B thermal reactions on this basis.

We first focus attention on thermal electrophilic aromatic substitution reactions. The following trends are noted:

a) The regioselectivity of these reactions is controlled by the HO electron density of the aromatic substrate [88a, b, 89].

b) The ortho/para ratio in most electrophilic aromatic substitutions of Ph–R [90], where R is a pi donor substituent, can be successfully rationalized on the basis of HO electron density control of regioselectivity.

c) The overall rates of electrophilic aromatic substitution reactions can be correlated with the ionization potentials of the aromatic substrates. Typical data are shown in Table 28.

Table 28. Relative rates of bromination and chlorination of aromatic hydrocarbons

Aromatic substrate	IP^a (eV)	Relative rate of bromination in 85% HOAc[b]	Relative rate of chlorination in 85% HOAc[b]
(benzene)	9.2	1	1
(toluene)	8.9	650	340
(o-xylene)	8.6	5,300	2,030
(trimethylbenzene)	8.3	1,670,000	30,000,000
(tetramethylbenzene)	8.0	11,000,000	–
(pentamethylbenzene)	7.7	810,000,000	134,000,000

[a] IP values for benzene, toluene and o-xylene were taken from: Turner, D. W.: Advan. Phys. Org. Chem. *4*, 30 (1966). The other values were estimated.

[b] Olah, G. A.: Acc. Chem. Res. *4*, 240 (1971).

d) The dependence of regioselectivity on polarity follows our predictions. A beautiful example can be taken from the work of Olah and his collaborators [90]. The pertinent data shown in Table 29 clearly illustrate that, as reaction polarity is gradually increased, the para over ortho ratio tends to unity.

175

Table 29. The dependence of product distribution of electrophilic substitutions of toluene on the nature of the electrophile

Electrophile	2[para]/[ortho]
p$-$MeOC$_6$H$_4$CH$_2$Cl/TiCl$_4$	5.00
p$-$MeC$_6$H$_4$CH$_2$Cl/TiCl$_4$	4.17
p$-$HC$_6$H$_4$CH$_2$Cl/TiCl$_4$	2.70
p$-$ClC$_6$H$_4$CH$_2$Cl/TiCl$_4$	2.70
p$-$NO$_2$C$_6$H$_4$CH$_2$Cl/TiCl$_4$	1.15
p$-$MeOC$_6$H$_4$SO$_2$Cl/AlCl$_3$	33.30
p$-$MeC$_6$H$_4$SO$_2$Cl/AlCl$_3$	12.50
p$-$HC$_6$H$_4$SO$_2$Cl/AlCl$_3$	4.35
p$-$ClC$_6$H$_4$SO$_2$Cl/AlCl$_3$	2.94
p$-$NO$_2$C$_6$H$_4$SO$_2$Cl/AlCl$_3$	1.61

e) The regioselectivity, the polarity effect on rate and probably the selectivity-polarity relationship in thermal electrophilic aromatic substitutions could be discussed without having to define whether a type A or a type B ionic mechanism is operative. In a type A mechanism, the transition state arises from the crossing of the DA and D$^+$A$^-$ diabatic surfaces and precedes N$_\pi^*$ complex formation. On the other hand, in a type B mechanism, the transition state occurs past the N$_\pi^*$ complex preceding N$_\sigma^*$ complex formation and results from the interaction of the DA and D$^+$A$^-$ diabatic surfaces away from the crossing point. Since both mechanisms respond similarly to the various mechanistic tests discussed above, we envision no easy way of differentiating between type A and type B. The best way of approaching this extremely difficult problem should involve an experimental test of the variation of the distance separating the two reactants at the transition state as a function of reaction polarity. If a type A mechanism is operative, the distance should increase as reaction polarity increases. A decrease in distance as reaction polarity increases would be indicative of a type B mechanism.

A greater challenge is posed by photoelectrophilic aromatic substitutions, assumed to be of P2″ type, where intuition cannot be used to extrapolate from preexisting data since these reactions have only recently attracted the attention of chemists. Indeed, the mechanism of aromatic photosubstitution reactions seem to be a recurrent topic of discussion in departmental seminars as well as international meetings. Despite all this frenetic activity, it is safe to say that no general theoretical framework is yet available for treating these reactions. Hopefully, the present work will fill this gap.

The photodeuteration of anisole is a particularly suitable reaction for illustrating our approach. The HO and LU electron densities of the model system phenol are shown in Appendix I. Clearly, the HO electron densities vary in the order p > o > m while the LU electron densities vary in an opposite manner, i.e., m ≥ o > p. When such a pattern obtains, an unambiguous prediction can be made. Specifically, we expect that the orientational preference of the photoelectrophilic deuteration of anisole will vary in the order m $\geq$ o > p, i.e., the order of increasing LU electron density and simultaneously decreasing HO electron density.

Experimental results are conflicting. It is found that the product distribution of deuterated anisole is m(8%), o(7.5%) and p(1%) [91]. However, dedeuteration results [92] suggest that the product distribution is approximately 6 : 1 : 2 for o, m, and p, respectively. Clearly, more work is needed to clarify the situation.

As a second example, let us consider the case of photoelectrophilic deuteration of naphthalene. Unlike the case of anisole, the HO and LU electron densities vary in the same direction, i.e., the α position has a higher electron density than the β position in both HO and LU. Accordingly, *an unambiguous prediction cannot be made here because simultaneous maximization of the $LU^D - LU^A$ and minimization of the $HO^D - LU^A$ matrix elements is impossible.* One has to decide whether traversing the barrier on the excited surface or crossing from excited to ground state surface is more crucial. Experimentally, it is found [91] that this reaction exhibits a preference for α attack. Thus, we are led to hypothesize that photoelectrophilic substitution reactions are barrier controlled reactions. Hence, we anticipate that the regioselectivity of photoelectrophilic aromatic substitution will, in general, be controlled by the electron density of the LU of the aromatic molecule.

Experimental data pertaining to photoelectrophilic aromatic substitution of some aromatic systems are shown in Table 30 and the calculated LU electron densities are shown in Appendix I.

Table 30. Products distribution of photodeutration reactions of aromatic hydrocarbons

Aromatic substrate	Electrophile	Products		References
	CF_3CO_2D		(Major)	(1)
	$MeCO_2D(H)$ o : p : m 2 : 1.4 : 1			(2)

(1) de Bie, D. A., Havinga, E.: Tetrahedron *21*, 2359 (1965).
(2) Spilane, W. J.: Tetrahedron *31*, 495 (1975).

177

At this point, an interesting result should be noted. Specifically, if the efficiencies of a type A thermal electrophilic substitution and that of its photochemical counterpart depend upon the heights of the corresponding barriers, it is predicted that the thermal reaction will be more regioselective because the transition state is tighter. The regioselectivity observed in the deuteration of naphthalene under thermal and photochemical conditions is a case in point. The thermal reaction is HO electron density controlled, the photochemical reaction is LU electron density controlled, and the electron densities of the α and β naphthalene positions are identical in HO and LU. Hence, any difference in the α over β regioselectivity ratio has to arise from the fact that the transition states of the two reactions occur at different intermolecular distances. Indeed it is observed experimentally that the regioselectivity of the photoreaction is much smaller than that of the thermal reaction.

$$\text{naphthalene} + D^+ \xrightarrow{h\nu} \quad 7 \quad : \quad 2$$

$$\text{naphthalene} + D^+ \xrightarrow{\Delta} \quad 12 \quad : \quad 1$$

Electrophilic aromatic substitutions were treated as bicentric reactions while they could very well be treated as reaction sequences involving multicentric attack to form the N_π^* complex and subsequent bicentric interaction to form the N_σ^* complex. The labels shown below will be used in the discussion of multicentric N_π^* complex formation. Table 31 compares the various predictions of the bicentric and multicentric models for the case of electrophilic aromatic substitution of monosubstituted benzene where the substituent is a pi electron donor, e.g., phenol. It is clear that both models lead to similar predictions.

Table 31. The regioselectivity of N_π^* and N_σ^* formation in the electrophilic aromatic substitution of phenol

Reaction type	Regiochemical condition	N_π^*		N_σ^*	
		Bicentric model	Multicentric model	Bicentric model	Multicentric model
Thermal	$HO^D - LU^A$	p > o > m	mp ~ io > om	p > o > m	p > o > m
Photo-chemical	$LU^D - LU^A$	o ~ m > p	io ~ pm > om	o ~ m > p	o ~ m > p
	$HO^D - LU^A(-)$	p < o < m	mp ~ io < om	p < o < m	p < o < m

io   om   mp

Before we depart the subject of electrophilic aromatic substitutions it should be pointed out that we restricted our attention to cases which involve expulsion of H^+ from the aromatic substrate, which sometimes ensures that the addition step is rate determining. The treatment can be extended to include ipso electrophilic substitution. For instance, if the collapse of the N_σ^* complex to product is assumed to be insensitive to the nature of the leaving group, the preference for thermal electrophilic attack on toluene is predicted to vary in the following order: para > ipso > ortho > meta. In general, we expect ipso attack to become prominent when attack on the other nuclear positions is unfavorable because the corresponding HO electron densities are very small and the splitting of the two highest occupied benzenoid MO's is large.

Turning our attention to thermal nucleophilic aromatic substitution reactions, the following can be said:

a) The regioselectivity of these reactions is controlled by the LU electron density of the aromatic substrate [88c, 89].

b) The overall rates of nucleophilic aromatic substitution reaction can be correlated with the electron affinities of the aromatic substrates. Typical data are shown in Table 32.

c) No systematic studies of the dependence of positional selectivity on the electronic nature of the reactants have yet been reported.

Table 32. Relative rates of nucleophilic aromatic substitutions

X	$\log k^a$	X	$E_a{}^b$	$\log A$
NH_2	−7.6	CF_3	31.5	11.0
Me	−5.3	COMe	29.2	11.2
H	−4.3	NO_2	24	11.2
Br	−3.25			
COOMe	−1.44			

[a] Bunnett, J. F., Zahler, R. E.: Chem. Revs. 49, 273 (1951).
[b] Miller, J., Yan, W. K.: J. Chem. Soc. 1963, 3492.

The mechanism of photonucleophilic aromatic substitutions, assumed to be P2″ photoreactions, has attracted considerable interest in recent years. As in the case of photoelectrophilic aromatic substitutions, the mechanisms of such reactions appear as a topic for discussion in international meetings with astounding frequency. Typical experimental data on the orientation preference of photonucleophilic aromatic substitutions are collected in Table 33 while the calculated HO and LU electron densities of the aromatic substrate are tabulated in Appendix I. Before proceeding with the discussion of the data certain cautionary remarks are in order.

As we have stated before, the preferred regioselectivity of nucleophilic photosubstitutions will be one which simultaneously maximizes the $HO^D - HO^A$ matrix element and minimizes the $HO^D - LU^A$ matrix element. By stating this selection rule we mean that the ideally preferred orientation will be the one which has the lowest photochemical barrier and can undergo the most efficient decay to the ground state. However, this requirement cannot be met in aromatic molecules where the HO and LU have identical electron density at any given center. This is the case of even alternant hydrocarbons where the pairing theorem holds [93]. Accordingly, we hypothesized that in photoelectrophilic aromatic substitution the product distribution is determined by the height of the photochemical barrier. In photonucleophilic aromatic substitution, this is controlled by the HO electron density of the aromatic.

Table 33. Regioselectivity of photonucleophilic aromatic substitutions

Aromatic substrate	Nucleophile	Products		References
OMe	$N^- = EtO^-, CN^-$	N		(1)
	$\overline{CN} + O_2$	OMe, CN 53%	OMe, CN 47%	(2)
NO₂, OMe	CN^-	NO₂, CN		(3)
	OH^-	NO₂, OH		(4)

Table 33 (continued)

Aromatic substrate	Nucleophile	Products	References
NO$_2$ / OPO$_3$	OH$^-$	NO$_2$ / O$^-$	(5)
OMe / NO$_2$	CN$^-$ + O$_2$	NC / OMe / NO$_2$ 29% + CN / OMe / NO$_2$ 26%	(6)
	OH$^-$	OH / NO$_2$ 31% + OMe / OH 3%	(7)
NO$_2$ / OMe	OH$^-$ + O$_2$	NO$_2$ / OH 20% + OH / OMe 80%	(8)
	CN$^-$ + O$_2$	OMe / CN / NO$_2$ + OMe / CN	(3, 6, 9)
2 3 1 4	CN$^-$	CN + NC	(10)

Table 33 (continued)

Aromatic substrate	Nucleophile	Products	References
	CN^-		(10)
	CN^-		(11)

(1) Barltrop, J. A., Bruce, N. J., Thompson, A.: J. Chem. Soc. (C) *1967*, 1142.
(2) Nilsson, S.: Acta Chem. Scand. *27*, 329 (1973).
(3) Letsinger, R. L., McCain, J. H.: J. Amer. Chem. Soc. *88*, 2884 (1966).
(4) a) Havinga, E.: K. Ned. Akad. Wet. Versl. Gewone Vergad, Afd. Naturk. *70*, 52 (1961);
     b) Havinga, E., de Jongh, R. O.: Bull. Soc. Chim. Belg. *71*, 803 (1962); c) de Jongh, R. O.,
     Havinga, E.: Rec. Trav. Chim. Pays-Bas *85*, 275 (1966); d) Havinga, E., de Jongh, R. O.,
     Kronenberg, M. E.: Helv. Chim. Acta *50*, 2550 (1967); e) den Heijer, J., Spee, T., de Gunst,
     G. P., Cornelisse, J.: Tetrahedron Let. *1973*, 1261.
(5) de Jongh, R. O., Havinga, E.: Rec. Trav. Chim. Pays-Bas *87*, 1318 (1968).
(6) Letsinger, R. L., McCain, J. H.: J. Amer. Chem. Soc. *91*, 6425 (1969).
(7) de Vries, S., Thesis, Leiden 1970.
(8) a) Letsinger, R. L., Steller, K. E.: Tetrahedron Let. *1969*, 1401; b) de Vries, S., Havinga, E.:
     Rec. Trav. Chim. Pays-Bas *84*, 601 (1965).
(9) Letsinger, R. L., Hautala, R. R.: Tetrahedron Let. *1969*, 4205.
(10) Vink, J. A. J., Lok, C. M., Cornelisse, J., Havinga, E.: Chem. Commun. *1972*, 710.
(11) Lok, C. M.: Thesis, University of Leiden, Leiden 1972.

Further complications may arise after decay of the excited complex to ground state N_σ^* complex. At this stage of the reaction, the yield of products will be determined by the leaving group ability [94, 95]. When the leaving group is a "poor" leaving group (e.g. H^-) and there are no appropriate conditions (a H^- sink) for its abstraction, the N_σ^* complex may decompose back to products. This complication may create a situation where the distribution of products will be similar to that encountered in a thermal reaction involving rate determining decomposition of the N_σ^* complex. This problem does not arise in photoelectrophilic aromatic substitution where H^+ can easily depart.

182

We now turn to the discussion of the experimental data (see also Appendix I). Examination of Table 33 reveals the following trends:

a) The photosubstitution reaction of anisole provides an intriguing example. Here, the C-1 and C-4 positions are the two positions of highest HO and lowest LU electron densities. In the presence of O_2 and protic solvent, which can assist the departure of H^-, nucleophilic attack can occur at positions C-2, C-3 and C-4. The order of HO electron densities is C-4 > C-2 > C-3 while the order of LU electron densities is the exact reverse. Accordingly, we can unambiguously predict that the nucleophile will attack these nuclear positions in the following order of increasing preference: C-4 > C-2 > C-3. This is found experimentally. In the absence of O_2, the superiority of MeO^- as a leaving group along with the fact that C-1 bears a high HO electron density and zero LU electron density dictates preferential nucleophilic attack at C-1. More data showing the ortho-para directing power of methoxy are presented in Table 34.

b) The photosubstitutions of m-methoxynitrobenzene and its phosphate analogue constitute interesting cases where the positions with the highest HO densities are C-6 and C-3 and C-3 is the carbon with the lowest LU density. This fact, together with the superiority of RO^- over H^- as a leaving group dictates substitution at C-3.

c) The photosubstitution of o-methoxynitrobenzene by CN^- in the presence of oxygen yields a main product which results from attack at the position of the highest HO density and the lowest LU density, i.e., at C-5. When O_2 is excluded, substitution by OH^- occurs preferentially at the site with the second highest HO density which bears the superior leaving group, MeO^-.

d) In p-methoxynitrobenzene the HO and LU densities vary in the order C-1 > C-4 > C-3, C-5 > C-2, C-6 and C-4 > C-1 > C-2, C-6 > C-3, C-5, respectively. With OH^- in the presence of O_2, substitution occurs preferentially at C-1 which has the highest HO density and relatively low LU density. The situation is similar with CN^- in the presence of O_2, although an additional product arises from attack at C-3, a position having high HO density and lowest LU density.

e) Table 33 reveals that biphenyl, naphthalene and phenanthrene give substitution at the site with the highest HO electron density. These sites also have the highest LU electron density. However, the following data indicate that the photosubstitution is controlled by the HO electron density. Thus, Letsinger et al. [96, 97] reported that 1-nitronaphthalene undergoes nucleophilic substitution at C-1 by CN^-, H^- and CH_3O^-, whereas 1-methoxynaphthalene undergoes substitution by CN^- mainly (98%) at C-4 [98]. In addition, 2-nitronaphthalene undergoes substitution at C-1 [99, 100] and the presence of O_2 is crucial for the substitution at this position. The difference between the substitution reactions of 1-nitronaphthalene and 1-methoxynaphthalene is in full agreement with the substituent effect on the HO electron density. Specifically, NO_2 causes a larger coefficient of C-1 and a smaller coefficient of C-4, whereas methoxy operates in the opposite direction. Finally, 1-nitro-4-methoxynaphthalene is substituted by a variety of nucleophiles at C-1 [96, 97, 101] which is in accord with the larger HO coefficient at this site. The same phenomenon is observed [102]

Table 34. Photonucleophilic substitution reactions of methoxy substituted aromatic compounds

Aromatic substrate	Nucleophile	Product	References
OMe / Cl (para)	X^- X = CN, OH, OMe	OMe / X (para)	(1)
OMe, OMe (ortho)	CN^-	OMe, CN (ortho)	(1)
OMe, OMe (meta)	CN^-	NC, OMe, OMe	(1)
MeO, OMe, OMe	CN^-	MeO, CN, OMe	(1)
OMe, OMe, OMe	CN^-	CN, OMe, OMe	(1)
MeO, OMe, OMe	CN^-	MeO, CN, OMe, OMe	(1)

(1) Cornelisse, J.: Pure Appl. Chem. 41, 433 (1975).

in the case of 4-nitrobiphenyl which is photosubstituted by CN^- at position 4 having the highest HO electron density.

g) Non alternant aromatic hydrocarbons are ideal systems for testing the validity of our selection rules. Azulene and its derivatives clearly demonstrate a thermal-photochemical dichotomy. As shown in Table 33, azulene is photosubstituted by CN^- at the position of highest HO and lowest LU density, whereas the thermal reaction results in substitution at positions 4 and 6 which are the sites with the highest LU density [103, 104]. 1-nitroazulene is also substituted at position 1 which has the highest HO electron density

At this point, the reader is warned that the multiplicity of the excited state reactant in the photoreactions collected in Tables 30, 33, and 34 and discussed in the text is in some cases unspecified, in some others singlet, and in still some others triplet. A further complication is the lack of definite information regarding the symmetry type of the active singlet or triplet excited state [105]. Finally, when the excited states of **D** or **A** are closely spaced (e.g., substituted benzenes) the relative ordering of DA* configurations may not necessarily be the same as that of A* states, i.e., the ZIDMOO LCFC approach may break down. Accordingly, we have assumed as a *working hypothesis* that the lowest singlet and triplet excited states of substituted aromatics arise from HO → LU electron promotions. With this assumption, the regioselection rules for optimization of the photochemical barrier are applicable to both singlet and triplet reactions. *A definitive identification of the reactive excited states of substituted aromatic molecules coupled with the methodology outlined here will go a long way in further refining the proposed regioselection rules.*

Finally, it must be emphasized that the reactions treated in this chapter were assumed to be ionic involving addition of an electrophilic ion pair having very high electron affinity or addition of a nucleophilic ion pair having very low ionization potential. However, solvent induced pseudoionic aromatic substitutions can also be envisioned. For example, a situation of the latter type may arise in the case of nucleophilic aromatic substitutions involving amine nucleophiles. The reader should convince himself that the same reactivity trends will obtain as when aromatic substitution is initiated by ionic addition.

13. Even-Odd Intermolecular Bicentric Reactions

13.1 Potential Energy Surfaces for Radical Additions

We now turn our attention to EO bicentric intermolecular reactions. Typical cases involve initial addition of a radical to an unsaturated system. The resulting adduct may then eliminate a chemical group giving rise to product formation.

Once again, the rates and regioselectivity of these reactions can be adequately understood by using a minimal basis set which includes the no bond configuration, the two lowest energy charge transfer configurations and the locally excited configuration where the excited partner is the unsaturated substrate.

We shall exemplify our approach by reference to a reaction where the radical acts as the donor and the unsaturated system as the acceptor. The basis set configurations and interaction matrix are shown in Fig. 52.

The equations of the diabatic surfaces are given below:

$$\epsilon(DA) \simeq S \tag{13.1}$$

$$\epsilon(D^+A^-) \simeq I_D - A_A - C + S' \tag{13.2}$$

$$\epsilon(D^-A^+) \simeq I_A - A_D - C' + S' \tag{13.3}$$

$$\epsilon(DA^*) \simeq {}^3G(HO^A \rightarrow LU^A) + S' \tag{13.4}$$

The P.E. surfaces are shown in Fig. 53 and the mechanistic details revealed by them can be expressed simply in terms of chemical equations:

186

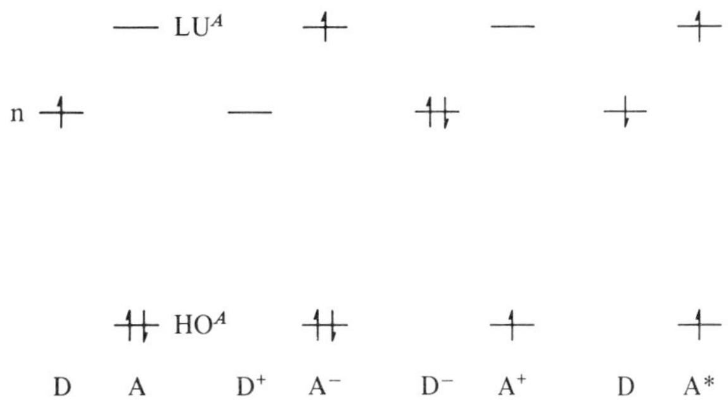

Fig. 52. Basis configurations and interaction matrix for nucleophilic radical addition to an unsaturated system

	DA	D^+A^-	D^-A^+	DA*
DA	—	$n-LU^A$	$n-HO^A$	0
D^+A^-		—	0	$n-HO^A$
D^-A^+			—	$n-LU^A$
DA*				—

Thermal: $\quad D + A \rightarrow (D \ldots A) \rightarrow N_\sigma \rightarrow$ Products

Photochemical: $\quad D + A^* \rightarrow (D \ldots A^*) \rightarrow$

$$D + A \quad\xrightarrow{-h\nu} M \xrightarrow{-h\nu} N_\sigma \rightarrow \text{Products}$$

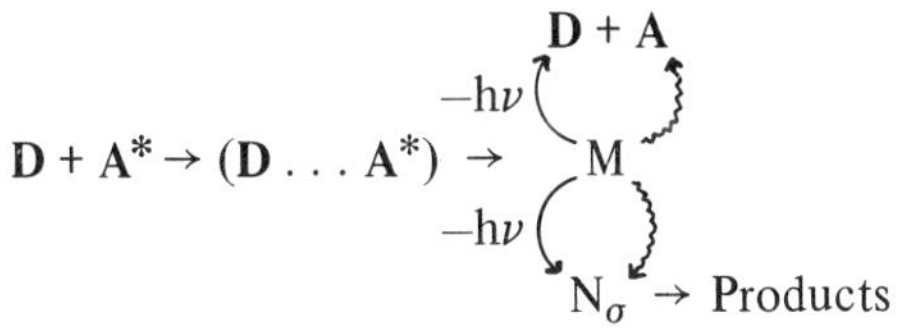

The following predictions can be made:

a) A thermal reaction will proceed in such a manner as to maximize the matrix element $\langle DA|\hat{P}|D^+A^-\rangle$ which is a $HO^D - LU^A$ matrix element.

b) A photochemical reaction will proceed in such a manner as to maximize the matrix element $\langle DA^*|\hat{P}|D^+A^-\rangle$, which is of the $HO^D - HO^A$ type, and minimize the $\langle DA|\hat{P}|D^+A^-\rangle$ matrix element, which is of the $HO^D - LU^A$ variety.

The effect of reaction polarity upon reaction rate and the reactivity-selectivity relationship can be discussed exactly as in the case of nonionic aromatic multicentric reactions.

A similar treatment can be given for reactions where the radical plays the role of the acceptor. The various predictions are summarized in Table 35.

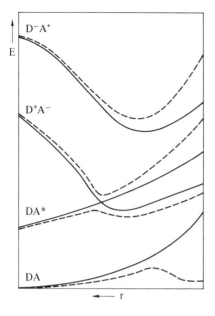

Fig. 53. Diabatic (solid lines) and adiabatic (dashed lines) surfaces for addition of a nucleophilic radical to an unsaturated system

Table 35. Reactivity aspects of radical even-odd additions

Model reaction	Reaction type	Regiochemical condition			Selectivity[a] response
		Ground surface	Excited surface	Decay process	
Nucleophilic radical addition	Thermal	$HO^D - LU^A$	–	–	↑
	Photochemical	–	$HO^D - HO^A$	$HO^D - LU^A(-)$	↓
Electrophilic radical addition	Thermal	$HO^D - LU^A$	–	–	↑
	Photochemical	–	$LU^D - LU^A$	$HO^D - LU^A(-)$	↓

[a] An arrow pointing downwards denotes a decrease in selectivity as polarity increases, and conversely.

13.2 Reactivity Trends of Radical Substitution Reactions

The P.E. surfaces for radical additions can serve as the basis for discussing radical substitution reactions insofar as the addition step is rate determining. In the analysis of the experimental data, the reader should bear in mind that the regioselectivity of most thermal radical substitutions is expected to be low because of the energetic proximity of the D^+A^- and D^-A^+ diabatic surfaces which enforce different regiochemical trends via their interaction with the DA diabatic surface.

The regioselectivity of radical addition to unsymmetrical double bonds has received a substantial amount of attention. It has been demonstrated [106] that CF_3, CCl_3 and C_3F_7 radicals attack the molecules shown below at the β carbon.

$$
\begin{array}{cc}
\beta & \alpha \\
\downarrow & \downarrow
\end{array}
$$

$$CH_2=CHF$$
$$CH_2=CF_2$$
$$CHF=CF_2$$
$$CH_2=CHCl$$
$$CHCl=CCl_2$$
$$CF_2=CFCl$$
$$CF_2=CCl_2$$

In all cases, the radical plays the role of the acceptor and the olefin the role of the donor. For example, the I–A values for the reaction of CF_3 and $CH_2=CF_2$ are as follows:

$$I\,(CH_2=CF_2) - A\,(CF_3) \simeq 7.8 \text{ eV} \tag{13.5}$$

$$I\,(CF_3) - A\,(CH_2=CF_2) \simeq 12 \text{ eV} \tag{13.6}$$

Although good experimental values for the electron affinities do not exist and extrapolated data had to be used, these numbers can be relied upon to signify that the radical acts as the acceptor.

Having established the roles of the reaction partners, we can predict that the regioselectivity of the attack will be controlled by the HO electron density of the olefin. In all of the above olefinic substrates, the β position has the highest HO electron density. Accordingly, the experimental data cited above is in accord with the theoretical predictions.

The regiochemistry of the addition of a radical to a substituted benzene has been extensively studied. The experimental results can be easily understood. In each case, a brief discussion of the specific theoretical predictions is presented in order to make the subsequent analysis of experimental data unambiguous.

Consider two typical substituted benzenes: one bearing a pi electron acceptor, NO_2, and the second a pi donor group, OH. The HO and LU electron densities of these two molecules are given in Appendix I and the following specific predictions can be made:

a) In the case of Ph–OH, if the Ar^+R^- lies lower in energy than the Ar^-R^+ configuration (Ar = aromatic substrate, R = radical), the preferred regiochemistry will be controlled primarily by the HO electron density while the LU electron density will be exerting only a secondary control. Accordingly, the order of regiochemical preference will be para > ortho > meta.

The reaction of anisole and cyano radical is a case in point. The I–A values are as follows:

$$I\,(Ph-OMe) - A\,(CN) \simeq 5.4\ eV \qquad\qquad (13.7)$$

$$I\,(CN) - A\,(PhOMe) \simeq 14.5\ eV \qquad\qquad (13.8)$$

Typical experimental results are shown below, where F_x are partial rate factors [107]. The same pattern of orientational preference has been observed with the CH_2CN, CH_2COOH, and CF_3 radicals [108].

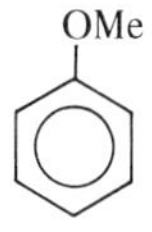

		F_o	F_m	F_p
(OMe-phenyl) + •CN		1.99	0.85	2.79

b) In the case of PhOH, if the $Ar^- R^+$ lies lower in energy than the $Ar^+ R^-$ configuration, the order of regiochemical preference will be ortho > meta > para. This order has not been achieved with any radical, at least in an unambiguous way.

c) In the case of $Ph-NO_2$, if the $Ar^- R^+$ lies lower in energy than the $Ar^+ R^-$ configuration, the order of regiochemical preference is predicted to be ortho > para > meta.

Typical experimental results are given below [109]:

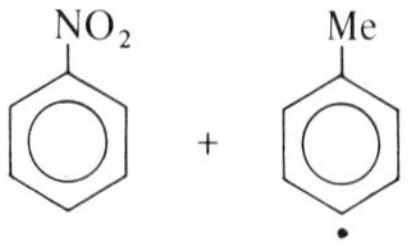

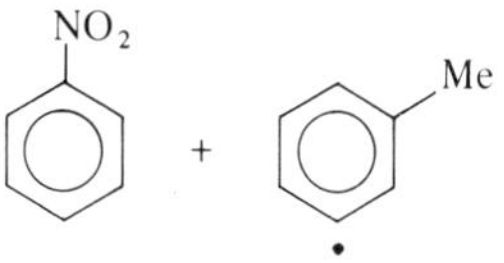

		F_o	F_m	F_p
(NO_2-phenyl) + (Me-phenyl radical)		6.1	1.2	5.8
(NO_2-phenyl) + (Me-phenyl radical)		5.5	1.2	4.7

The I−A values are estimated as follows:

$$I\,(Ph-NO_2) - A\,(Me-C_6H_4) \simeq 8.0\ eV \qquad\qquad (13.9)$$

$$I\,(Me-C_6H_4) - A\,(Ph-NO_2) \simeq 7.5\ eV \qquad\qquad (13.10)$$

d) In the case of $PhNO_2$, if the $Ar^+ R^-$ configuration lies lower in energy than the $Ar^- R^+$ configuration, weak or no regiochemical preference is predicted.

It should be noted that many additional experimental studies which illustrate the dependence of the orientation of radical additions to aromatics on the nature of the radical have been reported and the trends are in good agreement with the theoretical predictions [110].

Experimental evidence in support of the idea that an increase in reaction polarity will be accompanied by an increase in rate of radical addition to an unsaturated system abounds. Typical examples can be found in recent articles [106, 111].

In conclusion, the point should be made that the HO and LU electron densities of aromatic molecules and their derivatives can be estimated from the hydrogen hyperfine splitting constants of the corresponding radical cations and radical anions, respectively. Thus, Electron Spin Resonance (ESR) spectroscopy can become a tool for studying chemical reactivity. This procedure and related applications have been discussed in a full paper [112].

14. Odd-Odd Intermolecular Bicentric Reactions. Potential Energy Surfaces for Geometric Isomerization and Radical Combination

In this chapter, we shall discuss the reaction of neutral odd electron molecular species such as organic radicals to form a covalent or an ionic bond. However, before we embark on this mission, we shall briefly discuss as an example how bond formation occurs within a molecule.

We first consider the formation of a covalent bond from a singly occupied χ AO and a singly occupied ϕ AO. The key configurations and the static LCFC diagram are shown in Fig. 54. The following important trends should be noted:

a) The ground state of the bond has a dominant no bond contribution while the lowest excited state is primarily ionic.

b) The mixing of the no bond and lowest charge transfer configurations, and, hence, X–Y bond strength, is expected to increase as the ionization potential of one singly

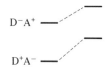

| D | A | D | A | D⁺ | A⁻ | D⁻ | A⁺ |

$$D \quad A \qquad D \quad A \qquad D^+ \quad A^- \quad D^- \quad A^+$$

Singlet   Triplet

D^-A^+ ──

D^+A^- ──

^3DA

^1DA

Fig. 54. Basis set configurations and static LCFC diagram describing the bond states for two singly occupied AO's centered on X and Y

occupied AO decreases, the electron affinity of the other singly occupied AO increases, and the corresponding interaction matrix element increases. This situation is illustrated by the greater bond dissociation energy of C–F as compared with that of C–O. However, in situations where an increase of the polarity of the system is counteracted by a "shrinkage" of the AO coefficients of the uniting centers, a greater reduction of the absolute magnitude of the interaction matrix element will dominate a smaller diminution of the energy gap separating the interacting DA and D^+A^- configurations. The smaller C–H bond dissociation energy of $PhCH_2$ –H as compared with that of $CH_3CH_2CH_2$ –H can be traced to such conflicting trends. The two contrasting cases are illustrated in Fig. 55. The comparison of $CH_3CH_2CH_2$ –H and Ph–CH_2 –H bond dissociation energies is instructive and merits attention.

Donor	Acceptor		$\sim I_D - A_A$ (eV)	$\langle DA|\hat{P}|D^+A^-\rangle$ (in k$\mathcal{S}$ units)
CH_3CH_2CH–H	$CH_3CH_2CH_2^{\cdot}$	H·	7.0	~1.00
Ph–CH_2–H	Ph–$CH_2^{\cdot}$	H·	6.5	0.76

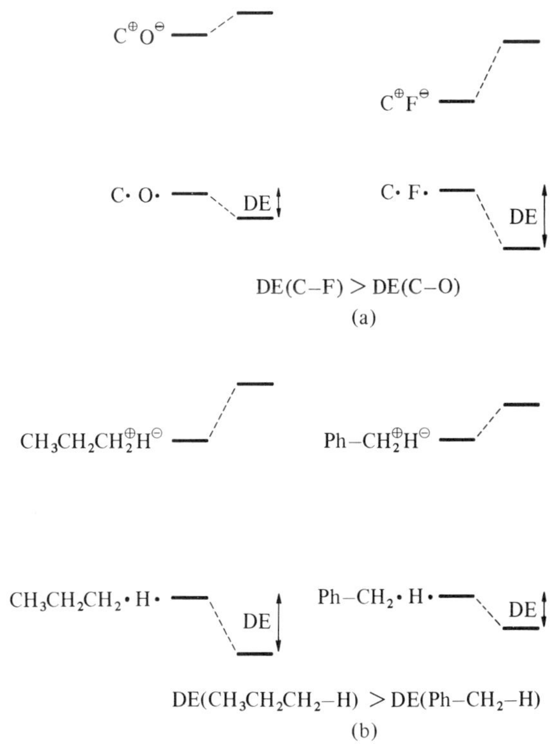

DE(C–F) > DE(C–O)

(a)

DE($CH_3CH_2CH_2$–H) > DE(Ph–CH_2–H)

(b)

Fig. 55. a) Energy gap and matrix element control of bond dissociation energy (DE). b) Matrix element control of bond dissociation energy (DE)

Increased delocalization in D leads to increased polarity coupled with an orbital electron density "dilution" effect. Hence, the polarity effect favors a stronger C—H bond in Ph—CH_2—H while the associated matrix element effect favors a stronger C—H bond in $CH_3CH_2CH_2$—H. When spatial orbital overlap is strong, matrix element control is imposed and the result is a smaller bond dissociation energy in the case of Ph—CH_2—H. This reflects the orbital electron density "dilution" effect which, in turn, is connected to pi conjugation in the formal Ph—CH_2 radical.

c) The above description is applicable to any covalent sigma or pi bond. An ionic bond will be formed when the D^+A^- configuration drops below the DA configuration. Obviously, this will occur when the singly occupied χ has a very low ionization potential and the singly occupied ϕ a substantial electron affinity, e.g., NaCl.

Let us now consider the simplest model OO reaction, namely, rotation about the X=Y bond in $\rangle$X=Y$\langle$. The diabatic surfaces and the resulting adiabatic surfaces are shown in Fig. 56. As the molecule rotates from 0° to 90°, the interaction matrix element which couples DA with D^+A^- and D^-A^+ decreases giving rise to a maximum on the ground surface and a minimum on the two excited surfaces. When the polarity of the X=Y bond becomes large, the D^+A^- surface will approach closely the DA surface. Accordingly, if the molecule is placed in a polar solvent, stabilization of the ionic surfaces will take place and this may lead to the interpenetration of the Φ_1 and Φ_2 adiabatic surfaces. The latter will create a high lying 90° intermediate on the ground adiabatic surface [113].

The above analysis suggests that *the barrier for rotation about a double bond will be influenced predominantly by the strength of the corresponding pi bond* since the transition state for the transformation, i.e., the 90° form, is not stabilized. Accordingly, substituent effects on geometric isomerization barriers can be understood by

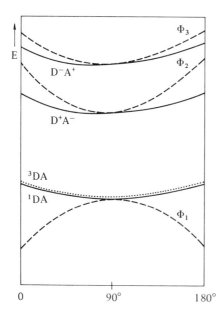

Fig. 56. Diabatic (solid lines) and adiabatic (dashed lines) singlet surfaces for bond rotation in $\rangle$X=Y$\langle$. Triplet adiabatic surface is indicated by dots.

realizing that the effect of substitution can be twofold, i.e., it may increase the polarity of the system and decrease the absolute magnitude of the $\langle DA|\hat{P}|D^+A^-\rangle$ interaction matrix element.

Interaction matrix element control is ensured when a smaller change of the energy gap of two diabatic surfaces is accompanied by a larger change of the corresponding interaction matrix element. This is expected when the distance between the two interacting fragments or molecules is large and spatial AO overlap is large. The effect of substituents on the activation energy for $\rangle C=C\langle$ bond rotation is consistent with matrix element control of the strength of the pi bond, a result similar to that obtained previously with regards to the effect of substituents on sigma bond strength.

Substrate	Activation energy (eV) (C=C Rotation)	References
CHD=CHD (cis)	65.0	[114a]
CHMe=CHMe (cis)	62.4	[114b]
CHCl=CHCl (trans)	41.9	[114c]

Next, we shall consider the surface manifold for the reaction of two radicals to form a sigma covalently bonded or an ionically bonded molecule. The symbols DA, D^+A^-, and D^-A^+ have their usual meaning and the diabatic and adiabatic surfaces which obtain in the two cases are shown in Figs. 57 and 58. Once again, polar solvents

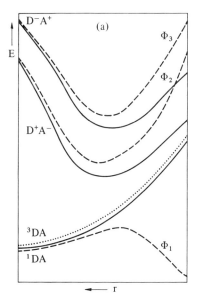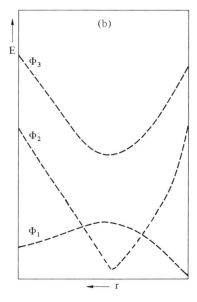

Fig. 57. a) Diabatic (solid lines) and adiabatic (dashed lines) describing the interaction of two atoms or radicals to form a bond. Triplet adiabatic surface is indicated by dots. b) The readjustment of the singlet adiabatic surfaces under the influence of polar solvent, e.g., solvolytic reaction

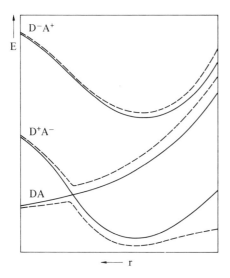

Fig. 58. Diabatic (solid lines) and adiabatic (dashed lines) surfaces describing the interaction of two atoms or radicals to form an ionic bond

may lead to a situation where a nonionic reaction is rendered ionic by virtue of the stabilization of the Φ_2 adiabatic surface which can now cross the Φ_1 adiabatic surface. Unimolecular solvolytic reactions are such cases.

In describing a simple radical combination reaction, one has to note that the steric function of the no bond, monoexcited, etc., diabatic surface equations will be very small simply because significant bond readjustment is not required, i.e., the two radicals may coalesce to form a product in *nearly* a frozen geometrical state. Accordingly, since the DA diabatic surface will be nearly flat, the activation energies for radical combination reactions are expected to be near zero. This is, indeed, what is observed experimentally [115].

The effect of reaction polarity on an intrinsically very low barrier height is expected to be very small. Indeed, it is found experimentally that most rate constants for alkyl radical combinations are within a factor of 10. Typical examples are collected in Table 36.

The above considerations lead us to anticipate that the most efficient way of enforcing significant reactivity differences is by affecting the steric function of the diabatic surface equations. This is possible by increasing the bulk of the combining radicals, i.e., steric effects can be responsible for pronounced rate reductions in radical combinations. Experimental data which suggests that this is so are available [116].

In the case of ionic atom combination reactions, the crossing of DA and D^+A^- diabatic surfaces occurs very early on the reaction coordinate and the barrier is practically zero. This situation is typical of the union of Na and Cl to form Na^+Cl^-.

The treatment of radical combination reactions presented above made use of a limited basis set of diabatic surfaces. Expansion of the basis set leads to novel insights. Thus, inclusion of locally excited configurations suggests that the energy of the excited ion pair intermediate is a function of the mixing of the configurations shown below.

196

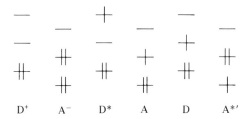

It is apparent that the stability of an ion pair X^+Y^- does not depend only on the energy of the D^+A^- configuration. In fact, it is expected to increase as the energies of the singly occupied → lowest unoccupied transition of $X\cdot$ and the highest doubly occupied → singly occupied transition in $Y\cdot$ decrease and the mixing of the three configurations shown above increases. This has the concomitant effect of reducing the polarity of the ion pair, i.e., reducing solvation. Whether differential electronic stabilization or differential solvation is more important is the topic of current investigations. A complete reworking of the concepts related to heterolytic cleavage reactions, such as solvolysis, seems necessary.

Table 36. Termination rate constants for alkyl radicals in solution[a]

Radical	Solvent	Approximate relative rates	$I-A$ (eV)
$H-\overset{\cdot}{C}H_2$	Cyclohexane	1.0	8.8
$Me-\overset{\cdot}{C}H_2$	Ethane	0.7	8.1
$Et-\overset{\cdot}{C}H_2$	Cyclohexane	0.4	7.3
$Ph-\overset{\cdot}{C}H_2$	Benzene	0.4	6.0
$(Me)_2\overset{\cdot}{C}-H$	Benzene	1.0	7.3
$(Me)_2\overset{\cdot}{C}-CN$	Benzene	0.1	7.3
$(Me)_2\overset{\cdot}{C}-Ph$	Benzene	0.8	5.9
$(Me)_3\overset{\cdot}{C}$	Cyclohexane	1.0	7.0
$(Me)_3\overset{\cdot}{S}i$	Cyclohexane	1.0	–
$(Me)_3\overset{\cdot}{S}n$	Cyclohexane	1.4	–

[a] Nonhebel, D. C., Walton, J. C.: "Free Radical Chemistry". London: Cambridge University Press 1974.

15. Odd-Odd Intramolecular Multicentric Reactions

15.1 Potential Energy Surfaces for Sigmatropic Shifts

Sigmatropic shifts typify OO intramolecular multicentric reactions. A successful theory has to account satisfactorily for the following reaction features:

a) Stereoselectivity. For illustrative purposes, we can restrict our attention to two different types of reaction modes: shifts occuring by retention of configuration in both migrating group (MG) and migrating framework (MF) and shifts occuring by inversion of configuration in either MG or MF. Examples are shown below.

Retention	Inversion	Inversion
in MG and MF	in MG	in MF

b) Toposelectivity. Sigmatropic shifts are intriguing since, in most systems, more than one reaction pathways are available.

c) The effect of polarity on reaction rates. We shall illustrate our approach by reference to the model reaction shown below.

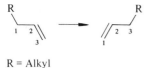

R = Alkyl

We shall restrict our attention to two stereochemical paths, one involving suprafacial migration with retention of configuration and the other suprafacial migration with inversion of configuration of the MG. These two different stereochemical paths will be designated SR and SI. The basis configurations are shown in Fig. 59. The equations of the diabatic surfaces are given below:

$$\epsilon(DA) \simeq S \tag{15.1}$$

$$\epsilon(D^+A^-) \simeq I_D - A_A + C + S' \tag{15.2}$$

$$\epsilon(D^-A^+) \simeq I_A - A_D + C' + S' \tag{15.3}$$

$$\epsilon(D^{*1}A) \simeq {}^2G(\phi_2 \rightarrow \phi_3) + S' \tag{15.4}$$

$$\epsilon(D^{*2}A) \simeq {}^2G(\phi_1 \rightarrow \phi_2) + S' \tag{15.5}$$

$$\epsilon(D^{+*}A^-) \simeq I_D - A_A + {}^2G(\phi_1 \rightarrow \phi_2) + C'' + S'' \tag{15.6}$$

$$\epsilon(D^{-*}A^+) \simeq I_A - A_D + {}^2G(\phi_2 \rightarrow \phi_3) + C''' + S'' \tag{15.7}$$

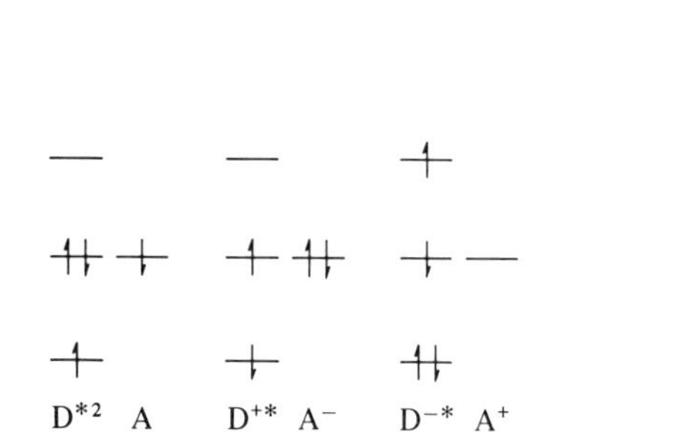

Fig. 59. Basis set configurations for the treatment of a 1,3 sigmatropic shift

The interaction matrix is given in Table 37. Unlike intermolecular reactions, the matrix elements change continuously as a function of the geometry of the reaction complex. Thus, for example, when the MG is near C_1 or C_3, only bicentric interactions need be considered and interaction matrix elements are large, unless a nodal plane mandates otherwise, *irrespective of the stereochemistry of the migration.* By

Table 37. Interaction matrix for 1,3 sigmatropic shift

	DA	D^+A^-	D^-A^+	$D^{*1}A$	$D^{*2}A$	$D^{+*}A^-$	$D^{-*}A^+$
DA	–	$(\phi_2 - \psi)$	$(\phi_2 - \psi)$	0	0	$(\phi_1 - \psi)$	$(\phi_3 - \psi)$
D^+A^-		–	0	$(\phi_3 - \psi)$	0	0	0
D^-A^+			–	0	$(\phi_1 - \psi)$	0	0
$D^{*1}A$				–	0	0	$(\phi_2 - \psi)$
$D^{*2}A$					–	$(\phi_2 - \psi)$	0
$D^{+*}A^-$						–	0
$D^{-*}A^+$							–

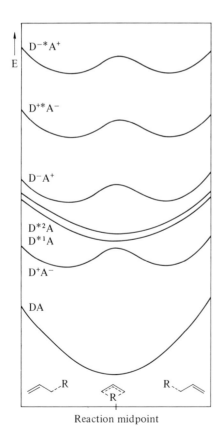

Reaction midpoint

Fig. 60. Diabatic surfaces for the treatment of a 1,3 sigmatropic shift

contrast, when the MG is on the midpoint of the C_1-C_3 side of the triangle defined by the three carbons of the MF, the interaction matrix elements become characteristic of each stereochemical path. Thus, it becomes apparent that near the beginning or end of the rearrangement process most diabatic surfaces will interact strongly, *irrespective of the stereochemical path.* By contrast, the strength of these interactions will critically depend upon the nature of the stereochemical path at the reaction midpoint.

The various diabatic surfaces for nonionic 1,3 sigmatropic shifts are shown in Fig. 60. An increase in polarity will alter the manifold in a manner which has been discussed before. By employing familiar reasoning, we reach the following conclusions:

a) P.E. surfaces for suprafacial 1,3 shifts proceeding by inversion can be constructed by using the DA, D^+A^- and $D^{*1}A$ diabatic surfaces (Fig. 61).

b) P.E. surfaces for suprafacial 1,3 shifts proceeding by retention can be constructed by using the DA, D^+A^-, $D^{+*}A^-$ and $D^{*1}A$ diabatic surfaces (Fig. 62).

The key features of the P.E. surfaces can be conveyed best by means of the chemical equations shown below, where [DA] stands for the unrearranged and [AD] for the rearranged molecule and the subscripts R and P refer to reactant-like and product-like intermediates, respectively.

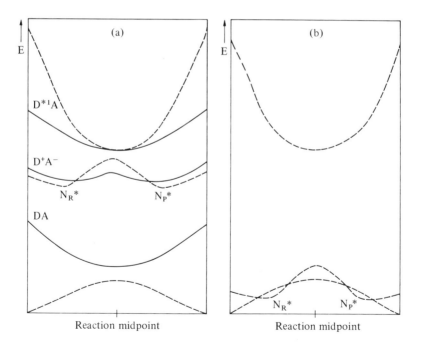

Fig. 61. a) Diabatic (solid lines) and adiabatic (dashed lines) surfaces for a 1,3 sigmatropic shift proceeding by inversion. b) The readjustment of the adiabatic surfaces under the influence of polar solvent

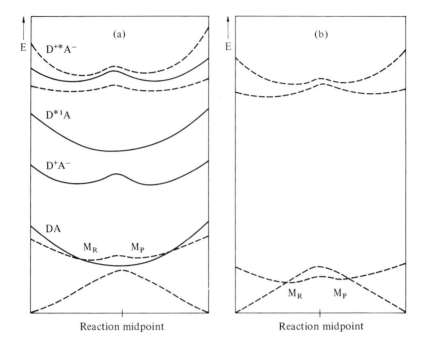

Fig. 62. a) Diabatic (solid lines) and adiabatic (dashed lines) surfaces for a 1,3 sigmatropic shift proceeding by retention. b) The readjustment of the adiabatic surfaces under the influence of polar solvent

a) Thermal 1,3 sigmatropic shift (SI path).

$$[DA] \xrightarrow[\text{Barrier}]{\text{Low}} [AD]$$

b) Photochemical 1,3 sigmatropic shift (SI path).

$$
\begin{array}{ccc}
[DA] & & [AD] \\
\updownarrow & \text{High} & \updownarrow \\
[DA]^* \longrightarrow N_R^* & \xrightarrow{\text{Barrier}} & N_P^* \longrightarrow [AD]^* \\
\downarrow & & \downarrow \\
[DA] + h\nu & & [AD] + h\nu'
\end{array}
$$

c) Thermal 1,3 sigmatropic shift (SR path).

$$[DA] \xrightarrow[\text{Barrier}]{\text{High}} [AD]$$

d) Photochemical 1,3 sigmatropic shift (SR path).

$$[DA]^* \longrightarrow M_R \xrightarrow[\text{Barrier}]{\substack{\text{Low}}} M_P \longrightarrow [AD]^*$$

$$[DA] \qquad [AD]$$

Note that the terms "high" and "low" barrier refer purely to electronic effects.

The key features of the P.E. surfaces can be described as follows:

a) The thermal barrier for the SI path is lower than the barrier for the SR path. This is due to the fact that the $DA-D^+A^-$ interaction is much stronger than the $DA-D^{+*}A^-$ interaction at the reaction midpoint. As a result, the thermal reaction will occur via the SI path, if electronic effects dominate steric effects. Both paths involve pericyclic bonding [117].

b) The photochemical SR path is more favorable than the photochemical SI path on two counts: a lower photochemical barrier and efficient decay to ground state reactants and products from the top of the photochemical barrier and ground state products from the M_P excited intermediate. The superiority of the photochemical SR path is due to the strong $D^+A^- - D^{*1}A$ interaction at the reaction midpoint, something absent in the case of the photochemical SI path. At this juncture, it should be mentioned that, if this interaction becomes very strong, the first excited P.E. surface may be modified in the sense that M_R, M_P and the barrier separating them may collapse to a single minimum from which decay to ground state reactants and products is very efficient. In such a case, the reaction mechanism will be as follows:

$$[DA]^* \longrightarrow M \longrightarrow [AD]^*$$

$$[DA] \qquad [AD]$$

The effect of polar solvent on the mechanism of [1,3] sigmatropic shifts is illustrated in Figs. 61 and 62 and is conveyed by means of the following equations:

a) Thermal 1,3 sigmatropic shift (SI path).

$$[DA] \longrightarrow N_R^* \longrightarrow N_P^* \longrightarrow [AD]$$

b) Photochemical 1,3 sigmatropic shift (SI path).

$$[DA]^* \rightsquigarrow N_R^* \longrightarrow N_P^* \longrightarrow [AD]$$

c) Thermal 1,3 sigmatropic shift (SR path).

$$[DA] \longrightarrow M_R \longrightarrow M_P \longrightarrow [AD]$$

d) Photochemical 1,3 sigmatropic shift (SR path).

$$[DA]^* \rightsquigarrow M_R \longrightarrow M_P \longrightarrow [AD]$$

A typical example of a sigmatropic shift occuring in a polar solvent is provided by the solvolysis of allylic halides. The elegant work of Sneen [118] has demonstrated the existence of two interconvertible unsymmetrical ion pair intermediates in the allylic rearrangement. These results are in good agreement with mechanistic predictions regarding this reaction based on the P.E. surfaces of Figs. 61 and 62.

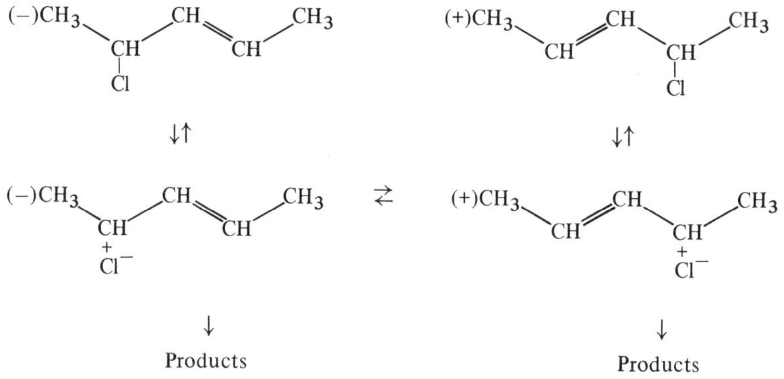

The toposelectivity of a sigmatropic shift occurring within one molecule and the relative rates of two sigmatropic shifts occurring in two different molecules are intimately related problems. For example, consider the two reactions shown below, each one assumed to occur via a SI stereochemical path. We further assume that in both cases the allyl MF acts as the acceptor partner and the MG as the donor partner, with

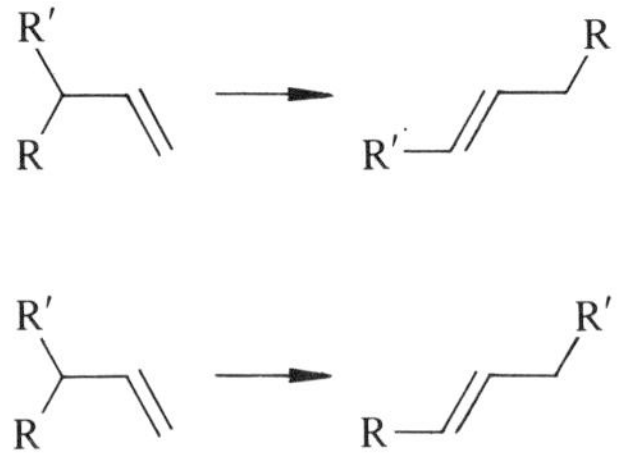

R′ being a better donor than R. In other words, we wish to compare the activation energies of two reactions, **D** + **A** and **D′** + **A**, where the latter has higher polarity. In doing so, one should consider the following:

a) The interaction of the DA and D^+A^- diabatic surfaces vs. that of the $D'A$ and D'^+A^- surfaces at the initial stage of the rearrangement, i.e., near covalent bonding of the two fragments. In this case, spatial overlap is very strong and differences in polarity are overcompensated by differences in matrix elements. If we assume that difference in polarity is imposed by the substituents borne by R and R', the operation of the orbital electron density "dilution" effect renders the matrix element controlled interaction of DA and D^+A^- superior to that of $D'A$ and D'^+A^-, i.e., weaker bonding accompanies greater polarity. As a result, a bond strength effect favors a more facile migration of R'.

b) The interaction of the DA and D^+A^- diabatic surfaces vs. that of the $D'A$ and D'^+A^- surfaces at the reaction midpoint. In this case, spacial overlap is weak and the effect of polarity becomes relatively more important. As a result, the superiority of the $DA-D^+A^-$ over the $D'A-D'^+A^-$ interaction is reduced. Thus, the more polar reaction will be faster.

The reader can verify that similar conclusions are valid if the two reactions occur via the SR stereochemical path.

On the basis of the above considerations, we can propose that $I_D - A_A$ will be an index of the activation energy of a sigmatropic shift with the reaction barrier becoming lower as $I_D - A_A$ decreases. Accordingly, relative rates as well as toposelectivity can be predicted in a simple manner.

A similar analysis can be given for photochemical shifts. In this case, excitation will be preferentially "localized" in the more polar of the C–R and C–R' bonds. If C–R' is more polar, R' will migrate in preference to R. Accordingly, the quantity $I_D - A_A$ may constitute a toposelectivity index of a photochemical sigmatropic shift.

The ideas presented in this section can be extended to any sigmatropic shift in the manner indicated below.

Number of electrons	Stereochemical path	P.E. surfaces
4N	SR	Fig. 62
4N	SI	Fig. 61
4N + 2	SR	Fig. 61
4N + 2	SI	Fig. 62

15.2 Reactivity Trends of Sigmatropic Shifts

Due to the fact that both retention and inversion paths involve pericyclic bonding, thermal nonionic 1,3 sigmatropic shifts can occur with complete inversion, or, if steric effects are prohibitive, complete retention. Experimental results bear out this prediction. For example, thermolysis of 1 has provided an example of a suprafacial sigmatropic shift proceeding with inversion [119].

On the other hand, reactions occuring by retention of the MG have also been discovered. Typical examples are shown below and more can be found in the literature [120, 121, 122].

Nonionic photochemical 1,3 sigmatropic shifts are expected to proceed with retention. One such shift has been reported by Cookson [121].

Examples of thermal sigmatropic shifts which exhibit lowering of the activation energy as the quantity I−A decreases are listed in Table 38. Only comparisons of systems belonging to the same class are permissible so that interaction matrix elements remain comparable and the polarity rule applicable. Other experimental trends consistent with our predictions are summarized below:

a) A 1,3 thermal shift of a carboxyl group across an allylic MF becomes faster as R_1-R_5 become more electron releasing and X more electronegative in 2 [123].

Table 38. Activation parameters of thermal sigmatropic shifts

Reactant	Thermal Product	I–A (eV)	E_a (kcal/mole)	log A (sec^{-1})	References
(structure: NC, RO$_2$C substituted diene)	(structure: NC, RO$_2$C product)	5.08	28.6	10.38	(1)
(structure: O-vinyl)	(structure: O pyran)	5.28	30.6	11.70	(2)
(structure: methyl cyclohexadiene)	(structure)	5.61	34.2	10.39	(3)
(structure)	(structure)	5.61	35.3	10.39	(3)
(structure)	(structure)	6.7	32.5	10.80	(4)
(structure: CD$_2$)	(structure: CD$_2$H)	7	$\Delta H^{\neq} =$ 35.4	–	(5)
(structure: CD$_3$)	(structure: CDH$_2$ CD$_2$)	7	37.7	11.86	(5)
(structure: cyclopentadiene)	(structures +)	7.7	20.4	–	(6)
(structure)	(structure)	7.7	19.9	–	(7)
(structure: D$_5$, H)	(structure: H, D$_5$)	7.9	24.3	12.11	(8)

Table 38 (continued)

Reactant	Thermal Product	I−A (eV)	E_a (kcal/mole)	log A (sec^{-1})	References
OMe (structure)	(structure) OMe	5.46	26.4	10	(9)
Ph (structure)	(structure) Ph	< 5.86	27.6	10.8	(10)
D (structure)	(structure) D	5.86	31.5	11.2	(11)

(1) Foster, E. G., Cope, A. C., Daniels, F.: J. Amer. Chem. Soc. *69*, 1893 (1947).
(2) Schuler, F. W., Murphy, G. W.: J. Amer. Chem. Soc. *72*, 3155 (1950).
(3) Frey, H. M., Solly, R. K.: Trans. Faraday. Soc. *64*, 1858 (1968).
(4) Frey, H. M., Pope, B. M.: J. Chem. Soc. (A) *1966*, 1701.
(5) Roth, W. R., König, J.: Ann. *699*, 24 (1966).
(6) McLean, S., Haynes, P.: Tetrahedron *21*, 2329 (1965).
(7) McLean, S., Webster, C. J., Rutherford, R. J. D.: Can. J. Chem. *47*, 1555 (1969).
(8) Roth, W. R.: Tetrahedron Let. *1964*, 1009.
(9) Nozoe, T., Takahashi, K.: Bull. Chem. Soc. Japan *38*, 665 (1965).
(10) Ter Borg, A. P., Kloosterziel, H.: Rec Trav. Chim. Pays-Bas *82*, 741 (1963).
(11) Ter Borg, A. P., Kloosterziel, H., Van Meurs, N.: Rec. Trav. Chim. Pays-Bas *82*, 717 (1963).

b) The reaction rate of *3* is $10^{10} - 10^{17}$ faster than that of *4* [124].

c) Reactivity patterns in cyclohexadienone chemistry are compatible with the predictions of our theory [125].

As we have seen before, our theoretical approach lends to the formulation of a simple topochemical rule which can be stated as follows: *a thermal sigmatropic shift will preferentially take place via the path which couples the best donor-acceptor frag-*

ments. In the space below, we provide a simple example of how this rule may be applied.

Consider an intramolecular 1,3 sigmatropic shift occuring in *5*.

The possible migration pathways are shown along with the quantities I—A, which constitute measures of the donor-acceptor relationship between MF and MG. For each migration mode, one has to calculate two values of I—A, one assuming that the MF is the donor and the MG the acceptor, and one for the reverse situation. The preferred pathway is *a* because it involves the lowest value of I—A, i.e., this pathway couples donor and acceptor fragments in the best possible way. Following similar reasoning, we predict that *5* will react faster than the unsubstituted derivative which can only undergo a 1,3 shift via a path akin to the inferior path *b*. Both these predictions are confirmed by experiment [126].

In Table 39, we provide examples of the application of the topochemical rule to systems which have been studied experimentally. These data are in agreement with the prediction that a sigmatropic shift will occur via the path which involves the best donor-acceptor combination.

The topochemistry of photochemical sigmatropic shifts proceeding in an electronically favored mode can also be predicted by reference to the index I — A. Table 40 contains examples of photochemical reactions which have been studied experimentally and which conform to the regiochemical rule. In each case, the preferred mode of migration is the one which couples the MF and MG in the best donoracceptor sense.

A recent review summarizes more pertinent data [127].

Table 39. Toposelectivity of thermal sigmatropic shifts

Compound	Possible products	I–A (eV)	Model MG	Model MF	Observed products
X = OMe, OAc (bicyclic structure with X)	A (structure)	5.6; 7.1	MeO$\overset{\bullet}{\text{C}}H_2$	(structure)	A[a]
	B (structure with X)	6.7; >7.1	CH$_3$–$\overset{\bullet}{\text{C}}H_2$		
(cyclopropyl OMe structure)	A (cyclopentene OMe structure)	5.9; ≥8.7	$\overset{\bullet}{\text{C}}H_3$	(structure OMe)	A[b]
	B (cyclopropyl OMe structure)	6.0; 11.0	$\overset{\bullet}{\text{O}}$Me	(structure)	
(diene structure)	A (structure)	5.6; 6.0	(structure)	(structure)	A[c]
	B (structure)	6.6; 7.7	$\overset{\bullet}{\text{C}}H_3$	(structure)	
Ph, NC, CN (structure)	A (structure CN, CN, Ph)	4.32; 7.2	Ph–$\overset{\bullet}{\text{C}}$(CH$_3$) H	(structure CN, CN)	A[d]
	B (structure CN, Ph, CN)	4.9; 11.6	$\overset{\bullet}{\text{C}}$N	(structure CN)	

[a] Scheidt, F., Kirmse, W.: Chem. Commun. *1972*, 716.
[b] See [126].
[c] Amano, A., Uchiyama, M.: J. Phys. Chem. *69*, 1278 (1965).
[d] Cookson, R. C., Kemp, J. E.: Chem. Commun. *1971*, 385.

Table 40. Toposelectivity of photochemical sigmatropic shifts

Reactant	Possible products		I−A (eV)	Model MG	Model MF	Preferred thermal product
	(A)		4.6	PhĊH₂		Aᵃ
	(B)		5.6	H₃C–ĊH₂		
	(C)		10.5	Ḣ		
	(D)		10.5	Ḣ		
	(A)		4.6	PhĊH₂		Aᵃ
	(B)		6.7	ĊH₃		
	(C)		10.5	Ḣ		
	(D)		10.5	Ḣ		

211

Table 40 (continued)

Reactant	Possible products	I−A (eV)	Model MG	Model MF	Preferred thermal product
	(A)	4.6	PhĊH$_2$		A^a
	(B)	10.5	Ḣ		
	(C)	10.5	Ḣ		
	(A)	4.3	Ph−ĊH CH$_3$		A^c
	(B)	10.5	Ḣ		
	(C)	10.5	Ḣ		

Table 40 (continued)

Reactant	Possible products	I–A (eV)	Model MG	Model MF	Preferred thermal product
	(A)	5.0			A[b]
	(B)	10.5	$\overset{\bullet}{H}$		
	(C)	5.6	$H_3C-\overset{\bullet}{C}H_2$		
	(D)	10.5	$\overset{\bullet}{H}$		
	(A)	5.2	$Me_2\overset{\bullet}{C}H$		A[d]
	(B)	7.2	$H_3\overset{\bullet}{C}$		
	(A)	4.4	$Ph_2\overset{\bullet}{C}H$		A[e]

Table 40 (continued)

Reactant	Possible products	I−A (eV)	Model MG	Model MF	Preferred thermal product
	(B)	5.2	$Ph_2\overset{\bullet}{C}H$		
	(A)	5.2	$H_3C-\overset{\bullet}{C}O$		A^f
	(B)	5.9	$CH_3-\overset{\bullet}{C}H_2$		
	(A)	6.7	$CH_3\overset{\bullet}{C}O$		A^g
	(B)	7.3	$\overset{\bullet}{C}H_3$		
	(C)	7.5	$\overset{\bullet}{H}$		

[a] Cookson, R. C., Hudec, J., Sharma, M.: Chem. Commun. *1971*, 107. Cookson, R. C., Hudec, J., Sharma, M.: Chem. Commun. *1971*, 108.
[b] Brown, R. F. C., Cookson, R. C., Hudec, J.: Tetrahedron *24*, 3955 (1968).
[c] Cookson, R. C., Kemp, J. E.: Chem. Commun. *1971*, 385.
[d] Hurst, J. J., Whitham, G. H.: J. Chem. Soc. *1960*, 2864.
[e] Boykin Jr., D. W., Lutz, R. E.: J. Amer. Chem. Soc. *86*, 5046 (1964).
[f] Cookson, R. C., Edwards, A. G., Hudec, J., Kingsland, M.: Chem. Commun. *1965*, 98.
[g] Baggiolini, E., Schaffner, K., Jeger, O.: Chem. Commun. *1969*, 1103.

214

16. Even-Even Intramolecular Multicentric Reactions

16.1 Potential Energy Surfaces for Ionic Rearrangements

A good understanding of ionic rearrangements can be obtained from an examination of the adiabatic P.E. surfaces which are generated incorporating the counterion wave-function in the zero order configurations which form the basis for the construction of the diabatic surfaces.

Our approach can be illustrated by reference to the model cationic shift shown below. The reaction is assumed to proceed by displacement of R by q accompanied

by displacement of X by $-q$ so that the motion of both R and X can be described by reference to a single coordinate $|q|$. In this reaction the MG is the R formal radical, the MF is the formal ethylene fragment, and X is the counteranion. The basis set configurations are shown in Fig. 63 and the interaction matrix in Table 41. For simplicity, diexcited configurations are not considered.

Table 41. Interaction matrix for a 1,2 cationic shift

	D_1D_2A	$D_1^+D_2A^-$	$D_1D_2^+A^-$	$D_1D_2^*A^-$	$D_1^+D_2^*A^-$	$D_1D_2^{+*}A$
D_1D_2A	—					
$D_1^+D_2A^-$	$n-n'$					
$D_1D_2^+A^-$	$HO-n'$	$n-HO$	—			
$D_1D_2^*A^-$	—	—	$n'-LU$	—		
$D_1^+D_2^*A^-$	—	—	$n-LU$	$n-n'$	—	
$D_1D_2^{+*}A^-$	—	—	—	$n'-HO$	$n-HO$	—

215

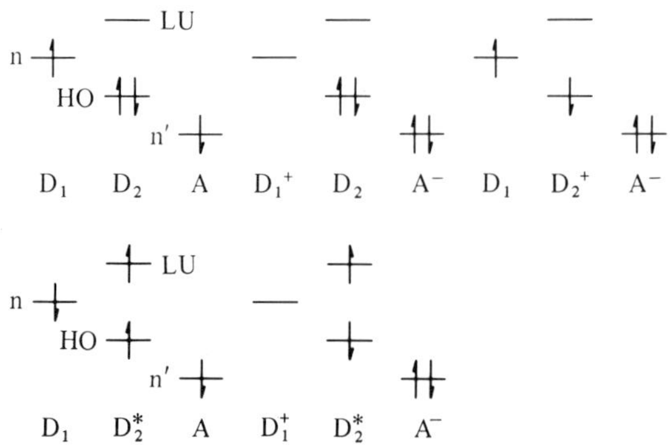

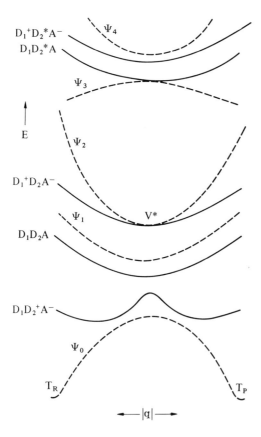

Fig. 63. Basis set configurations for a 1,2 cationic shift. D_1 is the migrating group, D_2 the migration framework and A the counteranion

The diabatic and adiabatic surfaces for suprafacial migration by retention as well as suprafacial migration by inversion of the R group are shown by Figs. 64 and 65. In both cases, it is assumed that X migrates by retention. The information of the final adiabatic P.E. surfaces can be conveyed by means of the chemical equations given below:

Fig. 64. Diabatic (solid lines) and adiabatic (dashed lines) surfaces for a 1,2 cationic shift proceeding suprafacially with inversion

a) Thermal SR Path (Φ_0 Surface)

$$T_R \xrightleftharpoons{\text{low barrier}} T_P$$

b) Photochemical SR Path (Φ_2 Surface)

$$T_R^* \rightarrow U^* \underset{}{\underleftarrow{}}\left\langle \boxed{\text{Inefficient } \Phi_2 \rightarrow \Phi_0 \text{ Decay}} \xrightarrow[\nearrow]{\searrow} \begin{array}{l} T_R \\ T_P \end{array}\right.$$

c) Thermal SI Path (Ψ_0 Surface)

$$T_R \xrightleftharpoons{\text{high barrier}} T_P$$

d) Photochemical SI Path (Ψ_2 Surface)

$$T_R^* \rightarrow V^* \left\langle \boxed{\text{Efficient } \Psi_2 \rightarrow \Psi_0 \text{ Decay}} \xrightarrow[\nearrow]{\searrow} \begin{array}{l} T_R \\ T_P \end{array}\right.$$

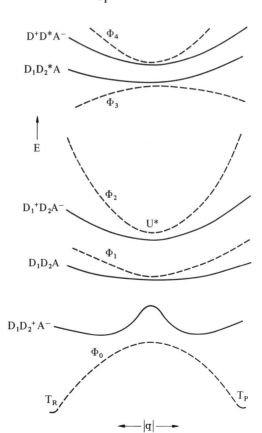

Fig. 65. Diabatic (solid lines) and adiabatic (dashed lines) surfaces for a 1,2 cationic shift proceeding suprafacially with retention

The terms "high", "low", "efficient", and "inefficient" are applicable only to relative comparisons. Furthermore, in the absence of quantitative information, the possibility exists that the double energy well diabatic surfaces, may actually involve a single energy well. The consequences are illustrated in Fig. 66.

At this point certain interesting aspects of the analysis presented above should be pointed out:

a) A cationic shift can be best described by using three fragments, D_1, D_2, and A. In this case, A has a very high electron affinity. By analogy, an anionic shift can be analyzed by using three fragments, D, A_1, and A_2, where D has a very low ionization potential.

b) The Φ_1, Φ_3, Ψ_1, and Ψ_3 adiabatic surfaces need not be considered when interested in the conventional migration of the MG. These surfaces describe the interaction of the MF and the counterion. Thus, if the assumed relative energy ordering of the diabatic surfaces is correct, such photoexcitation will not lead to rearrangement.

c) One can envision the following additional migration modes:
1. MG migrates by retention and X by inversion.
2. MG migrates by inversion and X by inversion.
These can be treated in the manner outlined before.

The major conclusions reached on the basis of the adiabatic surfaces can be summarized as follows:

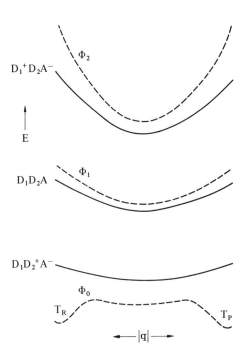

Fig. 66. Possible diabatic and adiabatic surfaces for a 1,2 cationic shift proceeding suprafacially with retention

218

a) The thermal reaction will proceed in a stereoselective SR manner because the SR path involves a lower barrier.

b) Both Φ_0 and Ψ_0 surfaces involve pericyclic bonding. Hence, if the SR path is blocked, a stereoselective SI mechanism can materialize.

c) Φ_0 as well as Ψ_0 have radical character.

d) The photochemical reaction will proceed in a stereoselective SI manner because the SI path involves a more efficient decay from excited to ground surface due to a smaller energy gap separating those two surfaces.

The analysis presented above can be generalized as indicated below:

Number of electrons	Type of shift	P.E. surfaces
4N	Cationic, SR	Fig. 64
	Cationic, SI	Fig. 65
4N	Anionic, SR	Fig. 64
	Anionic, SI	Fig. 65
4N + 2	Cationic, SR	Fig. 64
	Cationic, SI	Fig. 65
4N + 2	Anionic, SR	Fig. 64
	Anionic, SI	Fig. 65

The most important prediction is that a thermal ionic shift, where the electronically preferred mode is sterically unfavorable, can proceed via an alternative migratory mode in a stereoselective manner.

16.2 Reactivity Trends of Ionic Rearrangements

The conclusions of the above analysis are reminiscent of the ones reached in the case of sigmatropic shifts, where pericyclicity is imposed on two different types of reaction paths although one is favored more than the other. As an example, let us consider the cases of 1,2 cationic and 1,2 anionic shifts. In the former case, the SR path is favored electronically and sterically and both SR and SI paths involve pericyclic bonding. Thus, migration with retention is expected. In the case of a 1,2 anionic shift, the SI path is favored electronically over the SR path, while steric effects act in the opposite direction. Once again, pericyclic bonding is assured for both paths. Accordingly, in such reactions, one may obtain any result ranging from complete retention of the MG to partial retention due to competing SI and SR paths to complete inversion. In most experimental systems, steric effects are expected to play an important role. Accordingly, one can expect that Woodward-Hoffmann "forbidden" 1,2 anionic shifts will occur with a high degree of retention, at least in most cases.

Typical experimental observations of 1,2 anionic shifts are summarized in Table 42. The stereochemical results are in agreement with an expectation of SR stereoselectivity. In addition, the observation of CIDNP or ESR, in certain experimental systems is consistent with our analysis. Specifically, as we have already discussed, Ψ_0 (Fig. 64) has radical character and a shallow minimum on the top of the thermal barrier, which cannot be ruled out on the basis of our approach, may accomodate the intermediate which the organic chemist writes as a radical pair.

Table 42. Stereochemistry of 1,2 anionic shifts

Substrate	MG	CIDNP or ESR	Lowest Con- figuration	% Retention	References
$\overset{O}{\overset{\|}{Ph-C}}-\overset{\ominus}{CH}-\overset{\overset{CHMePh}{\|}}{\underset{\oplus}{N}(Me)_2}$	CHMePh	–	F^-G	97	(1)
$H_2C=CH-\overset{\ominus}{CH}-\overset{\overset{CHMePh}{\|}}{\underset{\oplus}{N}(Me)_2}$	CHMePh	–	F^-G	90	(2)
$Ph-\overset{\ominus}{CH}-\overset{\overset{CMeEtPh}{\|}}{O}$	CEtMePh	–	F^-G	80–90	(3)
$\overset{\ominus}{O}-\overset{\overset{CHDPh}{\|}}{\underset{\oplus}{N}(Me)_2}$	CHDPh	–	F^-G	80	(4)
$Ph-\overset{\ominus}{CH}-\overset{\overset{CHMeEt}{\|}}{O}$	CHEtMe	–	F^-G	62	(5)
$\overset{\ominus}{O}-\overset{\overset{CH_2Ph}{\|}}{\underset{\oplus}{N}(Me)_2}$	CH$_2$Ph	Yes	F^-G	–	(6)
		Yes	F^-G	–	(7)
$\overset{\ominus}{O}-\overset{\overset{tBu}{\|}}{C}HPh$	tBu	Yes	F^-G	–	(8)
$\overset{O}{\overset{\|}{Ph-C}}-\overset{\ominus}{CD}-\overset{\overset{CH_2Ph}{\|}}{\underset{\oplus}{N}(Me)_2}$	CH$_2$Ph	Yes	F^-G	–	(8)
$\overset{O}{\overset{\|}{Ph-C}}-\overset{\ominus}{CH}-\overset{\oplus}{S}Me$ $\underset{CH_2Ph}{\|}$	CH$_2$Ph	Yes	F^-G	–	(8)

(1) Brewster, J. H., Kline, M. W.: J. Am. Chem. Soc. 74, 5179 (1952).
(2) Jenny, E. F., Drury, J.: Angew. Chem. 74, 152 (1962).
(3) Schöllkopf, U., Schäfer, H.: Ann. Chem. 663, 22 (1963).
(4) Schöllkopf, U., Schäfer, H.: Ann. Chem. 683, 42 (1965).
(5) Schöllkopf, U., Fabian, W.: Ann. Chem. 642, 1 (1961).
(6) Schöllkopf, U., Ludwig, U., Patsch, M., Franken, W.: Ann. Chem. 703, 77 (1967).
(7) Lepley, A. R., Cook, P. M., Willard, G. F.: J. Am. Chem. Soc. 92, 1101 (1970).
(8) Schöllkopf, U.: Angew. Chem. Int. Edit. Engl. 9, 763 (1970).

17. Mechanisms of Electrocyclic Reactions

17.1 Introduction

An electrocyclic reaction is exemplified by the transformation shown below

We shall focus attention on the following aspects of these reactions:

a) Stereoselectivity. We need to consider three distinct stereochemical modes of reaction, namely, conrotation, disrotation, and monorotation. In the cases of disrotation and conrotation, there is a synchronous rotation of both ends undergoing union. On the other hand, in the case of monoration, only one end rotates initially and after or

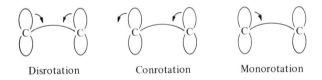

| Disrotation | Conrotation | Monorotation |

near completion of its rotatory motion the other end begins to rotate.

b) The effect of substituents upon reaction rate.

c) Toposelectivity. In certain types of molecules, there exist more than one electrocyclization pathways. An example is given below.

17.2 Reactivity Trends of Electrocyclic Reactions

Electrocyclic reactions can be viewed as intramolecular cycloadditions or cycloreversions. For example, the ring closures of butadiene and hexatriene can be viewed as intramolecular $2\pi + 2\pi$ and $4\pi + 2\pi$ cycloadditions, respectively. One can derive stereoselection rules for ring closures in a very simple manner by assuming that the two ethylenic fragments of butadiene can be treated as two isolated ethylenes and that the butadienic and ethylenic fragments of hexatriene can be treated as isolated butadiene and ethylene. Accordingly, one might expect that the effect of substitution upon the stereoselectivity of electrocyclic reactions will resemble the effect of substitution on the stereoselectivity of intermolecular cycloadditions. While this expectation is valid, one would have to note an important difference between the two aforementioned classes of reactions. Specifically, in electrocyclic reactions, steric effects are comparable for conrotation and disrotation. In intermolecular cycloadditions, steric effects are extremely different for s + s and s + a cycloaddend union. In the case of electrocyclic reactions, one cannot differentiate between LM and NLM paths since both conrotation and disrotation involve the same "amount" of nuclear motions. On the other hand, in the case of intermolecular cycloadditions, one can distinguish between LM and NLM paths, the former involving s + s and the latter s + a union of the cycloaddends. On the basis of these considerations, one can reasonably expect that the stereoselectivity of ring closures will primarily depend on electronic effects because the steric constraints imposed upon conrotation and disrotation are similar or can be made to be similar by appropriate design of the molecule to be investigated.

The P.E. surfaces for conrotatory ring closure of butadiene will be similar to those for $_2\pi_s + _2\pi_a$ intermolecular cycloaddition. The P.E. surfaces for disrotatory ring closure of butadiene will be similar to those for $_2\pi_s + _2\pi_s$ intermolecular cycloaddition. Finally, the P.E. surfaces for monorotation will bear close resemblance to those of TB union of two ethylenes. However, unlike intermolecular reactions, electrocyclic reactions commence at a "late stage". Thus, these reactions may not exhibit a photochemical barrier on the first excited surface.

In general, the ideas presented in the section dealing with $\pi + \pi$ cycloadditions can be extended to any $\pi + \pi$ electrocyclization in the manner indicated below.

Number of Pi electrons	Stereochemical path	P.E. surfaces
4N	Conrotation	Fig. 28
4N	Disrotation	Fig. 27
4N + 2	Conrotation	Fig. 28
4N + 2	Disrotation	Fig. 27

Experimental evidence, summarized in various papers and monographs [3, 4a, b, d–f] is in accord with the theoretical predictions concerning the stereoselectivity of thermal nonionic electrocyclizatione. The ionic counterparts have not yet been studied systematically.

The effect of reaction polarity on the relative rates of thermal electrocyclizations occurring via an aromatic path is in accord with the theoretical predictions. The relative rates of conrotatory ring openings of the molecules shown below are typical examples [128].

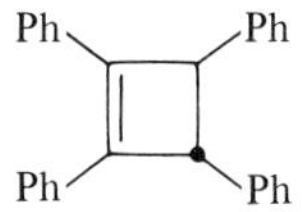

| E_A (kcal/mol) | 32.5 | 21.0 |

The predicted effects of reaction polarity and unsymmetrical substitution upon the rates of thermal nonionic electrocyclizations constrained to occur via an antiaromatic path are in agreement with the available experimental data. Thus, it was found [129] that *1* is thermally stable, whereas *2* opens to the seven-membered ring upon heating.

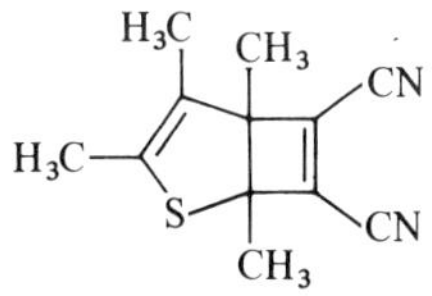

A host of reactions, which are thought to proceed *via* unsymmetrically substituted cyclobutenes, yield ring opened products despite the fact that the cyclobutene intermediate, e.g., is constrained to open by disrotation [130]. See also [131].

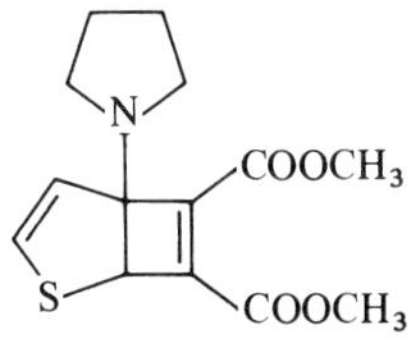

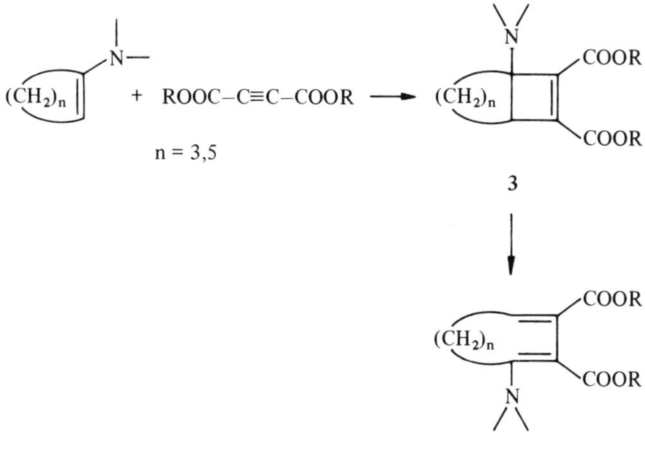

223

The enthalpies of activation for disrotatory ring opening of three typical Dewar-benzenes have been determined to be as follows [132a]:

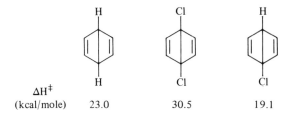

$\Delta H^{\ddagger}$ (kcal/mole)	23.0	30.5	19.1

Similar trends have been found in thermal ionic electrocyclization. The energy barriers for disrotation of constrained cationic species are much lower than those of the corresponding neutral molecules. Thus, $\Delta F^{\ddagger}$ for *4* is about 7 kcal/mole and 5 kcal/mole higher than $\Delta F^{\ddagger}$ for *5* and *6* respectively [132b].

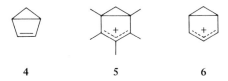

4	5	6

The effect of enhanced conjugation on thermal electrocyclizations constrained to proceed *via* antiaromatic paths can be exemplified by the facile valence isomerizations of *7* and *8* *via* conrotation and disrotation, respectively [133].

7   8

Turning our attention to photoelectrocyclizations, the stereochemistry of photochemical nonionic electrocyclic reactions has been thoroughly investigated and the results are in accordance with the theoretical predictions. The stereochemistry of ionic photoelectrocyclic reactions awaits experimental scrutiny.

Photoelectrocyclizations involving more than four pi electrons can occur *via* topochemically distinct paths. By using the same rationale as in the cases of intramolecular photocycloadditions, we predict that the preferred pathway will be the one which involves the best donor-acceptor combination. An example is given below and more can be found in the literature[1].

---

1   For a detailed discussion of photoelectrocyclizations, see: Epiotis, N. D., Shaik, S., Zander, W.: J. Am. Chem. Soc., submitted for publication.

Path *1* involves the inferior donor-acceptor combination of fragments B and B
(I − A = 12.3 eV). Path *2* is preferred since it involves the superior donor-acceptor
combination of fragments A and B (I − A = 10.7 eV).

18. Triplet Reactivity[1]

18.1 Introduction

Up until this point, we have been concerned almost exclusively with singlet thermal and photochemical reactions. We now turn our attention to triplet reactions which are different from the corresponding singlet reactions in the following two respects:

a) The diabatic surface energetic interrelationships are different, e.g., the energies of the $^1DA^*$ and $^3DA^*$ diabatic surfaces are different. This can become responsible for different photochemical barrier heights in singlet and triplet reactions.

b) In the case of a triplet reaction, a spin inversion process is required for decay to ground state reactants or products. Such a process is not demanded in a singlet reaction. We shall now consider the two key differences in some detail.

18.2 Excited State Spin Multiplicity and Photochemical Barrier Heights

We shall illustrate our approach by considering the singlet and triplet photochemical reactions of **D** and **A*** described by the diabatic surfaces DA, $^1D^+A^-$, $^3D^+A^-$, $^1DA^*$ and $^3DA^*$. It is assumed that the triplet locally excited diabatic surface has significantly lower energy than the singlet locally excited diabatic surface while the triplet and singlet charge transfer diabatic surfaces have comparable energies at long intermolecular distances.

The most commonly occurring patterns of interactions of the various diabatic surfaces are shown in Fig. 67. On the basis of previous discussions, we conclude the following:

---

1 For a more detailed development of the key equations and concepts, see: Shaik, S.: Ph. D. Dissertation, University of Washington 1978. A preliminary summary of the results was presented at the 173rd American Chemical Society National Meeting, New Orleans 1977, and the 32nd American Chemical Society Northwest Regional Meeting, Portland 1977.

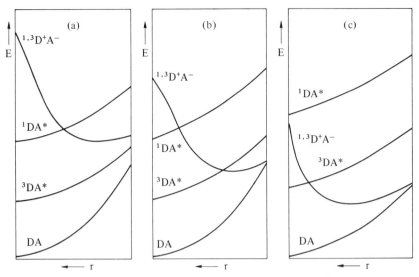

Fig. 67. Interrelationship of singlet and triplet diabatic surfaces in nonionic reactions. Stable singlet M excited intermediate can arise in a), b), and c). No triplet M intermediate can arise in a), an unstable one can be produced in b), and a stable one can be formed in c)

Case a). This situation may arise in reactions of very low polarity where the singlet reaction will proceed with a low barrier and ^1M intermediate formation. The triplet reaction will have a high barrier and will be inefficient.

Case b). This situation may arise in reactions of moderate polarity. The singlet barrier will be lower than the triplet barrier because $^1DA*$ is crossed earlier by $^1D^+A^-$ than $^3DA*$ by $^3D^+A^-$. Exceptions will arise when the interaction matrix elements are substantially different. In both singlet and triplet photoreactions, an excited intermediate M will be formed and ^1M will have a more negative heat of formation than ^3M. This situation is the one expected to be encountered most often in problems of chemical reactivity.

Case c). This situation may arise in reactions of substantial polarity. Here, the singlet reaction will have a high barrier and will be inefficient. The triplet reaction will proceed with a low barrier and ^3M intermediate formation. Stated in different language, case a) amounts to a singlet P2 vs. a triplet P1 photoreaction, case b) to a singlet vs. a triplet P2 photoreaction and case c) to a singlet P3 vs. a triplet P2 photoreaction.

The concept outlined above can be illuminated by reference to a particular example. Thus, case a) is typified by the triplet photochemical ring closure of butadiene to cyclobutene. As predicted, this reaction is known to be inefficient via the triplet state and *ab initio* computations show that the triplet state surface is characterized by a higher barrier [134]. Furthermore, the fact that triplet exciplexes are relatively rare whereas the number of singlet exciplexes is considerable, is in keeping with the fact that most organic photoreactions constitute case a) or case b) situations [56].

Next, we consider the mechanism of *radiationless* decay from a triplet to a singlet surface.

227

18.3 The Mechanism of Spin Inversion in Triplet Nonionic Photoaromatic Reactions

The LCFC treatment of the triplet $_2\pi_s + _2\pi_s$ photocycloaddition, which constitutes a model of a triplet nonionic photoaromatic reaction, proceeds in a manner similar to that of the corresponding singlet photoreactions, with the singlet diabatic surfaces comprising the Λ_2 packet replaced by their triplet counterparts. The course of the triplet photoreaction is described in the following mechanistic scheme:

$$D + {}^3A^* \rightleftarrows (D \ldots {}^3A^*) \rightleftarrows {}^3M \xrightarrow{\text{Spin inversion}} \text{Products}$$

Qualitatively, this mechanism is the same as the singlet photoreaction mechanism with the exception of the energy positioning of the $^3\Lambda_2$ boundary and the spin inversion step. The efficiency of spin inversion is related to the magnitude of the interaction between $^3\Lambda_2$ and $^1\Lambda_3$ boundaries past the "hole" defined by the $^1\Lambda_1 - {}^1\Lambda_3$ two electron avoided crossing. In all subsequent discussions, the symbol Λ denotes a packet boundary. Furthermore, *it is uniformly assumed that $^3\Lambda_2$ passes through the middle of the "hole"*.

In order to evaluate the magnitude of the interaction between $^3\Lambda_2$ and $^1\Lambda_3$ we have to include in the electronic Hamiltonian the appropriate Spin Orbit (SO) coupling terms. Accordingly, the $^3\Lambda_2 - {}^1\Lambda_3$ interaction is expressed by the integral, $\langle {}^3\Lambda_2 | \hat{H}_{El} + \hat{H}_{SO} | {}^1\Lambda_3 \rangle$ where $\hat{H}_{El}$ is the electronic Hamiltonian and $\hat{H}_{SO}$ is the SO coupling operator. Henceforth, we shall refer to the latter as the SO operator. Being part of the total operator, $\hat{H}_{SO}$ must belong to the totally symmetric representation of the molecular symmetry group. Consequently, one may use group theoretical techniques in order to evaluate qualitatively the $\langle {}^3\Lambda_2 | \hat{H}_{SO} | {}^1\Lambda_3 \rangle$ integral. The $_2\pi_s + _2\pi_s$ complex shown below belongs to the C_s point group. Accordingly, the $^3\Lambda_2$ state *space part* belongs to the A'' representation whereas the $^1\Lambda_3$ state *space part* belongs to the A' totally symmetric representation.

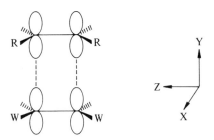

The total SO operator can be written as a double sum over the interactions of all electrons, i, with all nuclei, P, as shown in equation 18.1, where Z_P^* stands for the effective nuclear charge of nucleus P, r_{iP} stands for the distance between electron i

$$\hat{H}_{SO} = \sum_i \sum_P \frac{Z_P^* e^2}{2m^2 c^2} \frac{\hat{l}(i) \cdot \hat{s}(i)}{r_{iP}^3} \qquad (18.1)$$

and nucleus P and $\hat{l}(i)$ and $\hat{s}(i)$ are the orbital and spin angular momentum operations for electron i, respectively. Equation 18.1 can be rewritten as a sum of intra- and intermolecular contributions.

$$\hat{H}_{SO} = \hat{H}_{SO}^D + \hat{H}_{SO}^A + \hat{H}_{SO}' \qquad (18.2)$$

The first two terms on the right describe the intramolecular SO coupling contribution, while the third term describes the intermolecular contribution. The operators on the right hand of equation 18.2 can be expanded in terms of one electron operators as shown below:

$$\hat{H}_{SO}^D = \sum_i \hat{h}^D(i) \qquad (18.3)$$

$$\hat{H}_{SO}^A = \sum_i \hat{h}^A(i) \qquad (18.4)$$

$$\hat{H}_{SO}' = \sum_i \hat{h}(i) \qquad (18.5)$$

The expression for $\hat{h}(i)$ is given in equation 18.6:

$$\hat{h}(i) = \sum_P \frac{Z_P^* e^2}{2m^2 c^2} \frac{\hat{l}(i) \cdot \hat{s}(i)}{r_{iP}^3} \qquad (18.6)$$

In our discussions we shall focus attention on SO coupling effected via the intermolecular part of the SO operator since the corresponding interaction matrix element is connected with the stereochemistry of the triplet reaction.

We next consider the forms of the $^3\Lambda_2$ and $^1\Lambda_3$ wavefunctions. In the case of a nonionic photoaromatic reaction, the $^3\Lambda_2$ wavefunction can be approximated as shown below, where it is assumed that DA* is the lowest locally excited configuration.

$$^3\Lambda_2 \simeq N (a_1 \, ^3D^+A^- + a_2 \, ^3DA^*) \qquad (18.7)$$

The $^1\Lambda_2$ wavefunction can be approximated as follows:

$$^1\Lambda_2 = M (b_1 \, ^1D^*A^* + b_2 \, ^1D^{+*}A^- + b_3 \, ^1D^+A^{-*}) \qquad (18.8)$$

In treating matrix elements, one can make use of two equivalent formulisms, namely, one involving Cartesian operators and real atomic wavefunctions and another involving ladder operators and complex atomic wavefunctions. We shall make use of the former formulism since it is closely connected with the intuition of the organic chemist. In Cartesian space, the $\hat{l} \cdot \hat{s}$ operator is expanded as shown in equation 18.9, where the $\hat{l}$ components operate only on the orbital parts and the $\hat{s}$ components only on the spin parts of the wavefunctions. The effect of the $\hat{s}$ operators on spin wave-

$$\hat{l} \cdot \hat{s} = \hat{l}_x \hat{s}_x + \hat{l}_y \hat{s}_y + \hat{l}_z \hat{s}_z \tag{18.9}$$

functions and the effect of the $\hat{l}$ operators on atomic spatial wavefunctions can be gleaned by reference to Tables 43 and 44. In computing matrix elements with respect to the $\hat{l} \cdot \hat{s}$ operator, the following trends should be kept in mind:

a) The $\hat{s}_z$ operator leaves the α or β spin wavefunctions unaltered, i.e., an α or β spin wavefunction is an eigenfunction of $\hat{s}_z$. As a result, this operator will render zero any matrix element where the spin parts on the two sides of the operator do not match. By contrast, the $\hat{s}_x$ and $\hat{s}_y$ operators may have an opposite effect.

Table 43. Effect of spin momentum operators on atomic spin wavefunctions

Operator \ Spin Function	α	β
$\hat{s}_z$	$\dfrac{1}{2}\hbar\alpha$	$\dfrac{-1}{2}\hbar\beta$
$\hat{s}_x$	$\dfrac{1}{2}\hbar\beta$	$\dfrac{1}{2}\hbar\alpha$
$\hat{s}_y$	$\dfrac{i\hbar}{2}\beta$	$\dfrac{-i\hbar}{2}\alpha$

Table 44. Effect of angular momentum operators on Cartesian p atomic orbitals

Operator \ AO	p_z	p_x	p_y	
$\hat{l}_z$	0	$i\hbar p_y$	$-i\hbar p_x$	y
$\hat{l}_x$	$-i\hbar p_y$	0	$i\hbar p_z$	
$\hat{l}_y$	$i\hbar p_x$	$-i\hbar p_z$	0	

b) The effect of the $\hat{l}_k$ (k = x, y, z) operators on the real wavefunctions p_x, p_y, or p_z is multiplication by $i\hbar$ or $-i\hbar$ and rotation by 90° about the axis specified by the operator subscript.

The three components of the triplet $^3\Lambda_2$ wavefunction and the single component of the singlet $^1\Lambda_3$ wavefunction are shown below, where the notations $^{2S+1,M_s}\Lambda_2$ and $^{2S+1,M_s}\Lambda_3$ are being used.

$$^{3,-1}\Lambda_2 = N[a_1^{3,-1}(D^+A^-) + a_2^{3,-1}(DA^*)] \tag{18.10}$$

$$^{3,0}\Lambda_2 = N[a_1^{3,0}(D^+A^-) + a_2^{3,0}(DA^*)] \tag{18.11}$$

$$^{3,1}\Lambda_2 = N[a_1^{3,1}(D^+A^-) + a_2^{3,1}(DA^*)] \tag{18.12}$$

$$^{1,0}\Lambda_3 = M[b_1^{1,0}(D^*A^*) + b_2^{1,0}(D^{**}A^-) + b_3^{1,0}(D^+A^{-*})] \tag{18.13}$$

Using these wavefunctions we can compute, using well known procedures, the three components of SO coupling $\langle H'_{SO}\rangle_x$, $\langle H'_{SO}\rangle_y$ and $\langle H'_{SO}\rangle_z$. The selection rules derived from these integrals will now be discussed.

A. $\hat{l}_z$ Selection Rules

Group theoretical considerations indicate that $\langle \hat{H}'_{SO}\rangle_z$ vanishes in C_s symmetry. Accordingly, efficient SO coupling cannot be effected unless a molecular distortion destroys C_s symmetry and generates orthogonal AO's at the union site. In the problem at hand, $p_y \rightarrow p_x$ and/or $p_y \rightarrow -p_x$ AO rotations are required. Typical complexes which satisfy the latter condition are shown in Scheme 3 where the three pairs of complexes differ from each other in terms of the number and direction of AO rotations. By contrast, the two complexes within each pair differ from each other only in terms of the direction of rotations. In each case, the predicted *relative* SO coupling efficiency is stated.

Scheme 3 leads to the following selection rules:

a) Spin-orbit coupling is maximized by Perimolecular Conrotation or Endomolecular Disrotation. Both mechanisms are bisrotatory, i.e., they involve two 90° AO rotations.

b) Spin-orbit coupling is minimized by Perimolecular Disrotation or Endomolecular Conrotation. Both mechnisms are bisrotatory.

c) Intermediate spin-orbit coupling is produced by a monorotation mechanism.

Scheme 3. Rotational distortion motions and corresponding complexes for a $2\pi_s + 2\pi_s$ triplet photocycloaddition ($\hat{l}_z$ component)

$2\pi_s + 2\pi_s$ Geometry	Direction of AO rotations	Nomenclature	Resulting complex	SO Efficiency
	$p_{2y} \rightarrow p_{2x}$ $p_{3y} \rightarrow p_{3x}$	Perimolecular conrotation (PC)	(Ia)	Max.
	$p_{2y} \rightarrow p_{2x}$ $p_{3y} \rightarrow p_{3x}$	Perimolecular disrotation (PD)	(Ib)	Zero
	$p_{3y} \rightarrow p_{3x}$	Monorotation	(IIa)	Nonzero
	$p_{3y} \rightarrow p_{3x}$	Monorotation	(IIb)	Nonzero

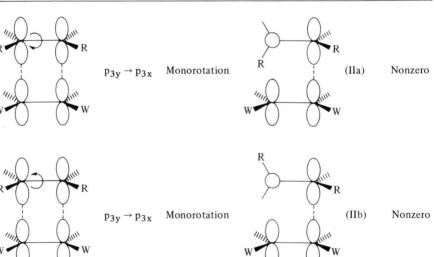

Scheme 3 (continued)

$2\pi_s + 2\pi_s$ Geometry	Direction of AO rotations	Nomenclature	Resulting complex		SO Efficiency
	$p_{3y} \to p_{3x}$ $p_{4y} \to p_{4x}$	Endomolecular conrotation (EC)		(IIIa)	Zero
	$p_{3y} \to p_{3x}$ $p_{4y} \to p_{4x}$	Endomolecular disrotation (ED)		(IIIb)	Max.

B. $\hat{I}_x$ Selection Rules

Group theoretical considerations indicate that $\langle \hat{H}'_{SO} \rangle_x$ vanishes unless a molecular distortion which preserves C_s symmetry and generates orthogonal AO's at the union sites is performed. The necessary AO rotations are $p_y \to p_z$ and/or $p_y \to -p_z$, as illustrated below. Clearly, a full 90° rotation is impossible. However, "partial rotation" is

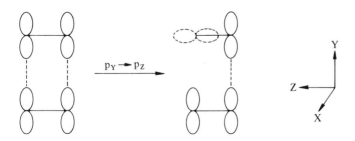

possible via pyramidalization of the uniting centers which mixes pi and sigma type MO's. A brief background discussion of pyramidalization is presented below.

In a molecule where pi-sigma separation can be defined, certain distortion modes may effect a mixing of pi and sigma type MO's. In our case, the distortion mode is pyramidalization. Its net effect amounts to mixing of π and σ bonding MO's as well as π^* and σ^* antibonding MO's of the same symmetry. The degree of sigma-pi mixing is proportional to the degree of pyramidalization. Accordingly, we define the new pi type MO's after pyramidalization as follows:

$$\pi = \left(\frac{1}{1+\lambda_1^2}\right)^{\frac{1}{2}} (\pi^0 + \lambda_1 \sigma) \tag{18.14}$$

$$\pi^* = \left(\frac{1}{1+\lambda_2^2}\right)^{\frac{1}{2}} (\pi^{0*} + \lambda_2 \sigma^*) \tag{18.15}$$

In the above expressions, π^0 and π^{0*} are the "pure" pi MO's of a planar olefin, π and π^* are the new pi type MO's after pyramidalization of the olefinic centers, and λ_1 and λ_2 are mixing coefficients. For a constant pyramidalization, the magnitude of λ_1 or λ_2 depends upon the energy separation and interaction matrix element of the σ^* and π^*, or, σ and π levels which are admixed by this molecular distortion.

In unraveling the pyramidalization modes which induce strong SO coupling, we are faced with problems similar to those encountered in unraveling the rotational modes having the same effect. Specifically, attention should be paid to number of pyramidalizations as well as the direction thereof. Typical complexes generated by pyramidalization of one or more olefinic centers are displayed in Scheme 4. The forms of the SO coupling matrix elements dictate the following trends:

a) As the number of pyramidalized centers increases, the absolute magnitude of SO coupling can increase. The pyramidalization mechanism which can best induce SO coupling is the one which involves two pairs of endomolecular pyramidalizations.

b) Within a group of complexes involving the same number of pyramidalized centers, one can define a maximally favorable and a minimally favorable pyramidalization mode with intermediate cases becoming possible as appropriate.

c) The effectiveness of the pyramidalization mechanism depends on the size of λ.

Scheme 4. Pyramidalization motions and corresponding complexes for a $_2\pi_s + _2\pi_s$ triplet photoaromatic cycloaddition ($\hat{l}_x$ component)

$_2\pi_s + _2\pi_s$ Geometry	Direction of AO rotations	Mechanistic type[a]	Resulting complex	SO Coupling efficiency
	$p_{4y} \rightarrow \lambda p_{4z}$	$+$		Nonzero
	$p_{2y} \rightarrow -\lambda p_{2z}$ $p_{4y} \rightarrow \lambda p_{4z}$	$+$ $-$		Max.
	$p_{2y} \rightarrow \lambda p_{2z}$ $p_{4y} \rightarrow \lambda p_{4z}$	$+$ $+$		Zero
	$p_{2y} \rightarrow -\lambda p_{2z}$ $p_{3y} \rightarrow -\lambda p_{3z}$	$+$ $-$		Max.

235

Scheme 4 (continued)

$2\pi_s + 2\pi_s$ Geometry	Direction of AO rotations	Mechanistic type[a]	Resulting complex	SO Coupling efficiency
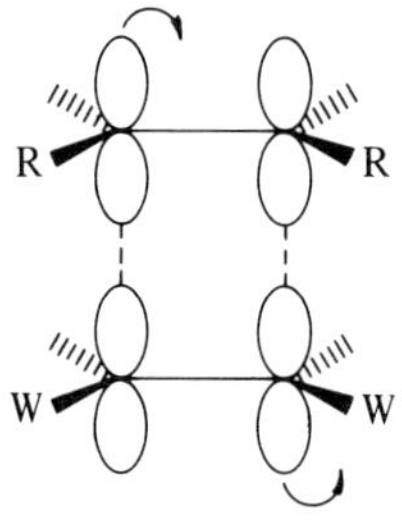	$p_{2y} \rightarrow \lambda p_{2z}$ $p_{3y} \rightarrow -\lambda p_{3z}$	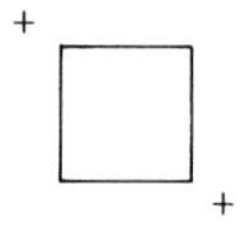	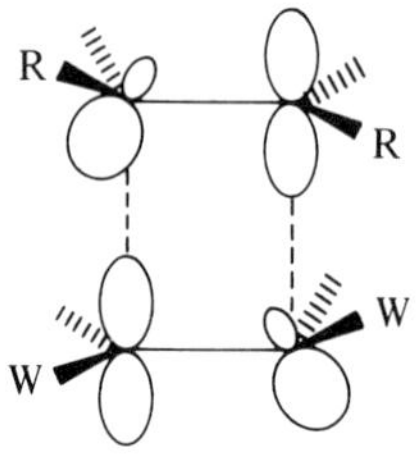	Zero
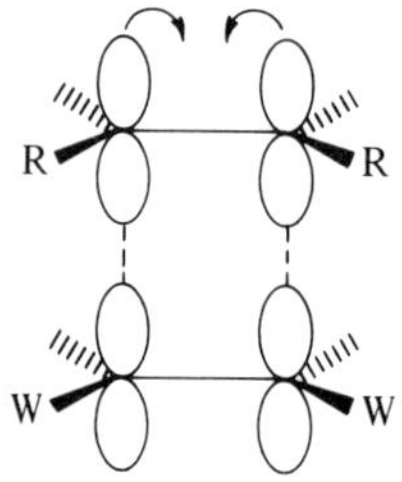	$p_{3y} \rightarrow -\lambda_3 p_{3z}$ $p_{4y} \rightarrow \lambda_4 p_{4z}$	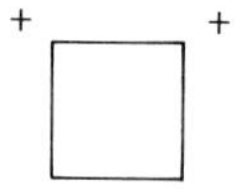	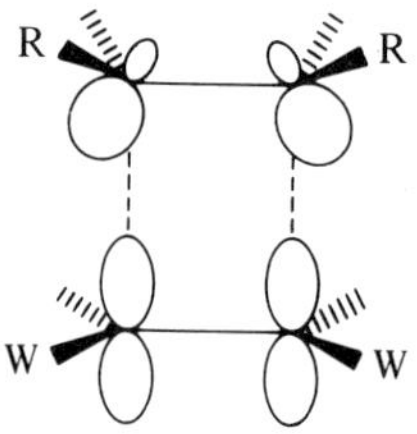	Max.
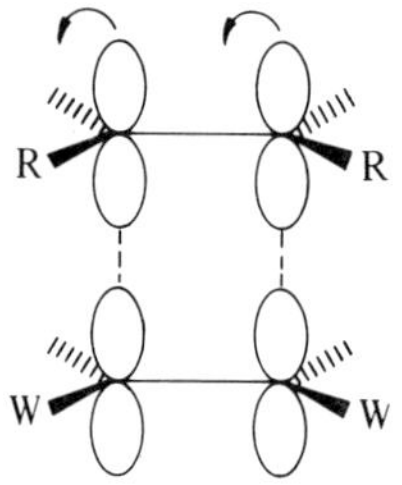	$p_{3y} \rightarrow \lambda p_{3z}$ $p_{4y} \rightarrow \lambda p_{4z}$	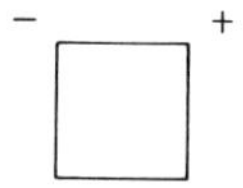	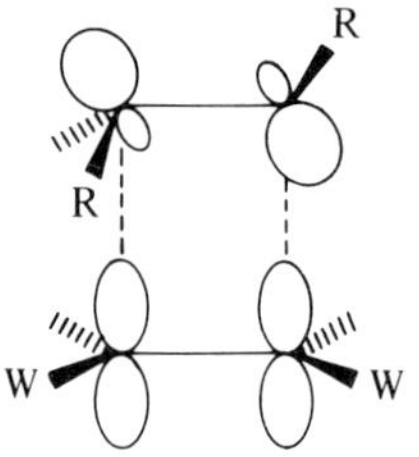	Zero
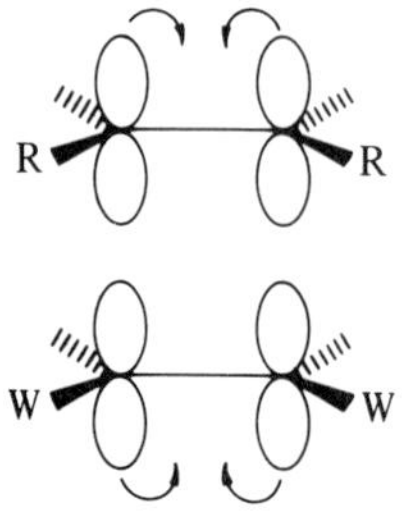	$p_{1y} \rightarrow \lambda p_{1z}$ $p_{2y} \rightarrow -\lambda p_{2z}$ $p_{3y} \rightarrow -\lambda p_{3z}$ $p_{4y} \rightarrow \lambda p_{4z}$	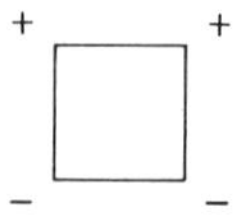	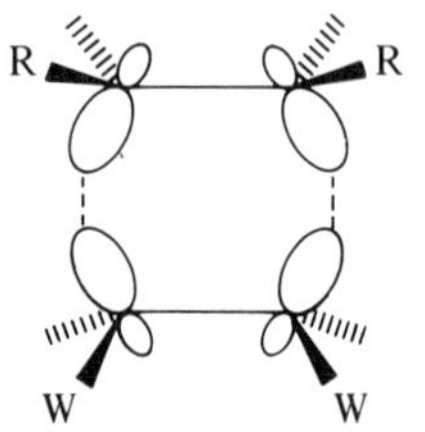	Max.

Scheme 4 (continued)

$2\pi_s + 2\pi_s$ Geometry	Direction of AO rotations	Mechanistic type[a]	Resulting complex	SO Coupling efficiency

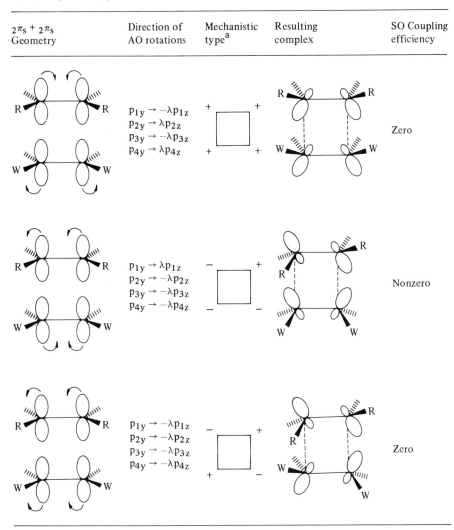

$$p_{1y} \rightarrow -\lambda p_{1z}$$
$$p_{2y} \rightarrow \lambda p_{2z}$$
$$p_{3y} \rightarrow -\lambda p_{3z}$$
$$p_{4y} \rightarrow \lambda p_{4z}$$

Zero

$$p_{1y} \rightarrow \lambda p_{1z}$$
$$p_{2y} \rightarrow -\lambda p_{2z}$$
$$p_{3y} \rightarrow -\lambda p_{3z}$$
$$p_{4y} \rightarrow -\lambda p_{4z}$$

Nonzero

$$p_{1y} \rightarrow -\lambda p_{1z}$$
$$p_{2y} \rightarrow -\lambda p_{2z}$$
$$p_{3y} \rightarrow -\lambda p_{3z}$$
$$p_{4y} \rightarrow -\lambda p_{4z}$$

Zero

[a] The signs denote the directions of the hybrid AO's, such that a plus (+) sign stands for a hybrid AO pointing downwards and a minus (−) sign stands for a hybrid AO pointing upwards.

C. $\hat{l}_y$ Selection Rules

By following similar reasoning as before, we conclude that the critical molecular distortion which can generate efficient SO coupling involves two types of AO rotations, namely, $p_y \rightarrow p_x$ and/or $p_y \rightarrow -p_x$ as well as $p_y \rightarrow p_z$ and/or $p_y \rightarrow -p_z$. As a result, the SO coupling matrix element with respect to the $\hat{l}_y$ operator dictates a combination of the rotational and pyramidalization mechanisms already discussed.

237

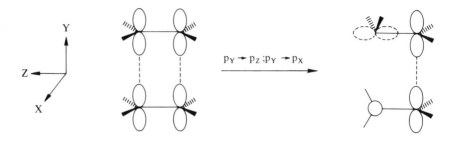

$$P_Y \rightarrow P_Z \, ; P_Y \rightarrow P_X$$

18.4 The Mechanism of Spin Inversion in Triplet Nonionic Photoantiaromatic Reactions

The course of the triplet $_4\pi_s + _2\pi_s$ photocycloaddition, which constitutes a model of a triplet nonionic photoantiaromatic reaction, can be conveyed by the following chemical equation:

$$\mathbf{D} + {}^3\mathbf{A}^* \rightarrow (\mathbf{D} \ldots {}^3\mathbf{A}^*) \rightarrow {}^3\mathbf{O} \xrightarrow{\text{Spin inversion}} \text{Products.}$$

The triplet photoreaction differs from the singlet photoreaction in the following important ways:

a) Whereas the singlet surface involves an antibonding contribution, due to ${}^1\Lambda_1 - {}^1\Lambda_2$ interpacket interactions, the triplet surface involves only bonding contribution. This arises because the ${}^1\Lambda_1 - {}^3\Lambda_2$ interaction is zero whereas the ${}^3\Lambda_2 - {}^3\Lambda_3$ interaction gives rise to bonding along the reaction sites and tends to minimize the photochemical barrier and the lowest excited tripletground singlet, ${}^3\Psi_1 - {}^1\Psi_0$ energy gap.

b) The ${}^3\Psi_1 \rightarrow {}^1\Psi_0$ radiationless decay of the triplet $_4\pi_s + _2\pi_s$ complex is accompanied by a spin inversion process.

   The efficiency of the spin inversion process depends on the size of the SO coupling matrix element $\langle {}^3\Psi_1 | \hat{H}_{SO} | {}^1\Psi_0 \rangle$. The magnitude of this integral can be evaluated qualitatively by group theoretical techniques. In the general case, the $_4\pi_s$

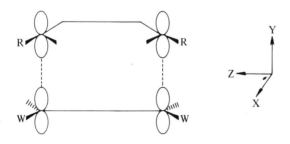

$+ _2\pi_s$ complex belongs to the C_s point group as illustrated below. Both $^3\Psi_1$ and $^1\Psi_0$ *space parts* transform as the totally symmetric irreducible representation, A'.

We shall now consider the conditions under which the crucial SO coupling matrix element is maximized. The approximate wavefunctions which describe the triplet excited and the ground surfaces are shown below:

$$^1\Psi_0 \simeq {}^1(DA) \tag{18.16}$$

$$^3\Psi_1 \simeq {}^3(D^+A^-) \tag{18.17}$$

Using these wavefunctions, we can compute the components of SO coupling and derive selection rules[1].

A. $\hat{I}_z$ Selection Rules

Group theoretical considerations indicate that $\langle \hat{H}'_{SO} \rangle_z$ is maximized by a molecular motion which preserves C_s symmetry and generates orthogonal AO's at the union sites. In the problem at hand, orthogonal $x-y$ AO relationships are required. Typical complexes which satisfy the latter condition and the associated SO matrix elements with respect to the $\hat{I}_z$ operator are shown in Scheme 5. The rotational mechanisms involving the lowest energetic price as well as the pyramidalization mechanism which maximizes SO coupling are shown below.

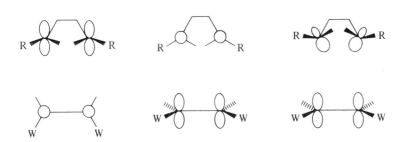

---

1 The efficiency of the spin inversion process depends also on the size of the SO coupling matrix element $\langle {}^3\Psi_1 | H'_{SO} | {}^1\Psi_1 \rangle$. By following familiar reasoning and using the approximate wavefunction shown below, we arrive at a set of equations which differ from those for $\langle {}^3\Psi_1 | H'_{SO} | {}^1\Psi_0 \rangle$ only in the sense that the MO interaction of the $HO^A - LU^D$ rather than the $HO^D - LU^A$ type.

$$^1\Psi_1 \propto M[b_1 {}^1(D^+A^-) + b_2 {}^1(D^*A^*)] \qquad (^1\Psi_1 = \text{lowest excited singlet})$$

Scheme 5. Rotational and pyramidalization mechanisms for a $4\pi_s + 2\pi_s$ triplet photoreaction ($\hat{l}_z$ component)

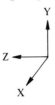

$4\pi_s + 2\pi_s$ Geometry	Nomenclature[a]	Resulting complex	SO Efficiency
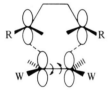	Endomolecular conrotation (EC)		Max.
	Endomolecular disrotation (ED)		Zero
	Endomolecular conrotation (EC)		Max.
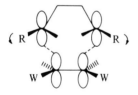	Monrotation	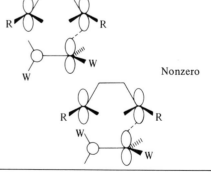	Nonzero

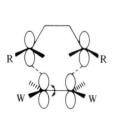

Scheme 5 (continued)

$4\pi_s + 2\pi_s$ Geometry	Nomenclature[a]	Resulting complex	SO Efficiency
	$+$ ⬡ $+$		Max.
	$+$ ⬡ $-$		Zero

[a] For an explanation of the sign convention, see footnote a in Scheme 2.

B. $\hat{I}_x$ Selection Rules

Group theoretical considerations indicate that $\langle\hat{H}'_{SO}\rangle_x$ is maximized by a molecular distortion which removes C_s symmetry and generates orthogonal AO's at the union sites. In the problem at hand, orthogonal $y-z$ relationships are required. The rotational and pyramidalization mechanisms which involve maximization of the SO coupling are shown below.

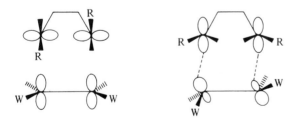

C. $\hat{I}_y$ Selection Rules

By following similar reasoning as before, we conclude that the critical molecular distortion which can generate efficient SO coupling involves $p_y \rightarrow p_x$ and/or $p_y \rightarrow -p_x$ as well as $p_y \rightarrow p_z$ and/or $p_y \rightarrow -p_z$ AO rotations. The rotational mechanism which maximizes the SO coupling and involves the least energetic price is shown below.

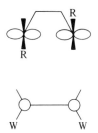

At this point it should be emphasized that the stereoselection rules for efficient SO coupling were based on the assumption that the $^3D^+A^-$ crosses the $^3DA^*$ (or $^3D^*A$) diabatic surface, i.e., the lowest energy diabatic surface of the $^3\Lambda_2$ packet for intermolecular distance of interest is assumed to be $^3D^+A^-$. This situation is very frequently encountered in organic triplet photoreactions. However, cases can be found where the relative energies of $^3D^+A^-$ and $^3DA^*$ (or $^3D^*A$) switch around. In such an event the crucial SO coupling matrix elements may, in principle, change and result in different spin inversion mechanisms. In photoantiaromatic reactions where $^3DA^*$ (or $^3D^*A$) lies below $^3D^+A^-$ at moderate or short intermolecular distances, the intermolecular SO coupling matrix element is of the $HO^D - HO^A$ or $LU^D - LU^A$ type. In such cases, the selection rules will resemble those for photoaromatic reactions.

18.5 Spin-Orbit Coupling Borrowing

Sor far we have considered only the intermolecular component of the total SO coupling matrix elements involved in triplet photochemical processes. We now illustrate how intramolecular contributions may arise and how they can influence the mechanisms of triplet photoreactions.

We shall illustrate our approach by considering the $_2\pi_s + _2\pi_s$ photoaromatic reaction. In this case, the intermolecular contribution to SO coupling involves $HO^D - LU^A$ and $HO^A - LU^D$ type matrix elements and these dictate the mechanisms discussed before. On the other hand, the intramolecular contribution involves $HO^D - LU^D$ and $HO^A - LU^A$ type matrix elements and these favor the maximum number of AO rotations possible. This intramolecular contribution may be termed SO coupling borrowing because it amounts to the rotational SO coupling contribution for the isolated triplet excited reactant.

The most important conclusion derived from the recognition of SO coupling borrowing is that the monorotation mechanism may be competitive with the ED mechanism in the case of $_2\pi_s + _2\pi_s$ cycloaddition. Similar considerations are applicable to the case of $_4\pi_s + _2\pi_s$ cycloaddition.

18.6 Mechanisms of Triplet Photoreactions

In discussing the mechanisms of different classes of photoreactions (i.e., photoaromatic, etc.), attention should be paid to the fact that, as one or both reactants suffer a geometrical distortion so that the requisite spin inversion process is optimized, the shapes of the adiabatic P.E. surfaces are simultaneously altered. Accordingly, we must seek a compromise geometry which is consistent with highly, though not maximally, efficient SO coupling as well as maintenance of bonding between the two reactants. This is a key premise for the subsequent discussions of triplet photoreaction mechanisms and it will be illuminated by reference to Scheme 6.

Consider a triplet photoaromatic reaction where the triplet photoexcited olefin approaches the ground singlet olefin in a $_2\pi_s + _2\pi_s$ manner (Structure I in Scheme 6). Past the "hole" defined by the $^1\Lambda_1 - ^1\Lambda_3$ avoided crossing (two-electron), ED converts structure I to structure II (Scheme 6). This process eliminates intermolecular bonding and simultaneously annihilates the "hole" because all photoaromatic P.E. surfaces are transformed to diabatic P.E. surfaces. The result is decay of the $_2\pi_s + _2\pi_s$ complex back to ground reactants, one or both of which are isomerized depending on the relative endorotatory efficiency of the olefins.

An alternative motion which can occur past the "hole" is endomolecular disrotation with simultaneous translation of the frame of one reactant relative to the other in the manner indicated by structure III (Scheme 6). This results in efficient SO coupling, generation of bonding between the reactants, and simultaneous annihilation of the "hole". This occurs because this mechanism transforms photoaromatic to photoantiaromatic P.E. surfaces. The result is that the triplet reaction complex decays to singlet ground products across an energy gap.

The situation becomes somewhat different when the rotational mechanism generates orthogonal AO's along one pair and nonorthogonal AO's along a second pair of union sites. Once again, consider the $_2\pi_s + _2\pi_s$ approach of a triplet excited and a ground olefin. Past the "hole", monorotation converts structure I to structure IV (Scheme 6). This process gives rise to efficient, though not maximal, SO coupling. In addition, it preserves intermolecular bonding but destroys the "hole" because all photoaromatic surfaces are transformed to nonaromatic P.E. surfaces. However, rotation by $\pm 90°$ can regenerate the "hole" and lead to efficient product formation.

The rotational mechanism of the $_4\pi_s + _2\pi_s$ cycloaddition can be illuminated in a similar manner. The basic conclusions are spelled out in Scheme 6.

Pyramidalization mechanisms stand in contrast to rotational mechanisms in the sense that pyramidalization distortion modes preserve the basic P.E. surface interrelationships present in the absence of distortion. Accordingly, although such mechanisms are inferior to the rotational mechanisms insofar as promoting efficient SO coupling, they retain the necessary features, e.g., intermolecular bonding, "hole", etc., for transformation of triplet excited reactants to ground singlet products.

At this point, it should be noted that the pyramidalization mechanism may become prominent under the following circumstances:

Scheme 6. SO coupling in $2\pi + 2\pi$ and $4\pi + 2\pi$ mechanisms

Complex[a]		Nomenclature	SO Coupling	Characteristic P.E. surface manifold	Stereochemical result
	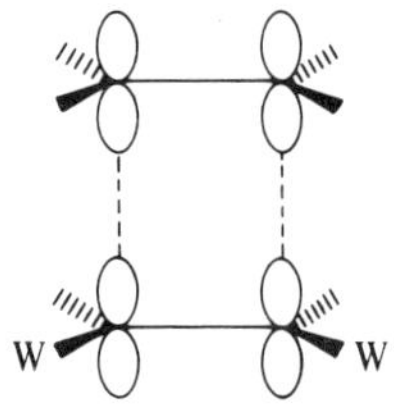 (I)	$2\pi_s + 2\pi_s$	Zero	Aromatic	–
	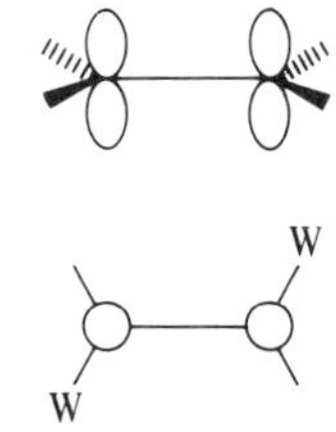 (II)	ED	Max.	Diabatic	Isomerization
	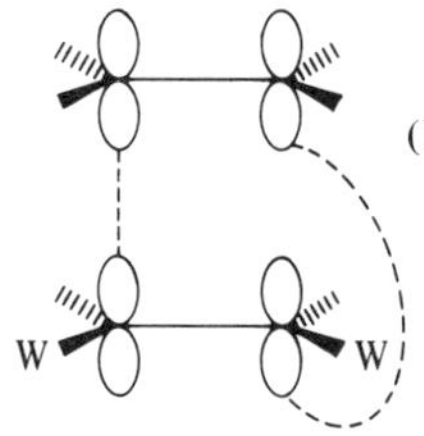 (III)	Translation rotation (ED)	Max.	Antiaromatic	2s + 2a Addition
	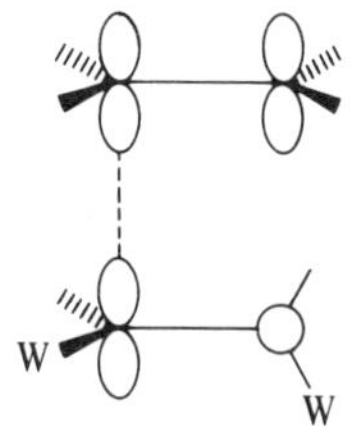 (IV)	Monorotation	Nonzero	Nonaromatic	Diradicaloid formation
	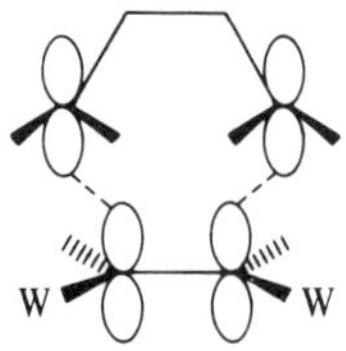 (V)	$4\pi_s + 2\pi_s$	Zero	Antiaromatic	–

Scheme 6 (continued)

Complex[a]	Nomenclature	SO Coupling	Characteristic P.E. surface manifold	Stereochemical result
(VI)				
	Translation rotation (EC)	Max.	Antiaromatic	4s + 2s Addition
(VII)				
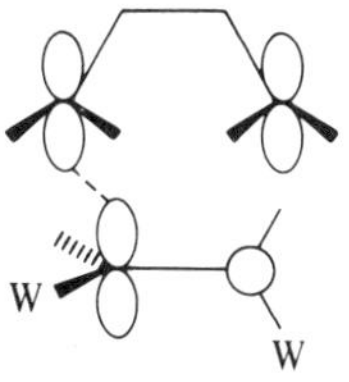 (VIII)	Monorotation	Nonzero	Nonaromatic	Diradicaloid formation

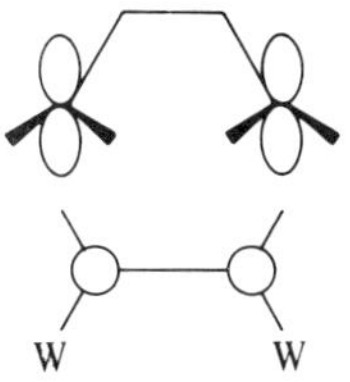

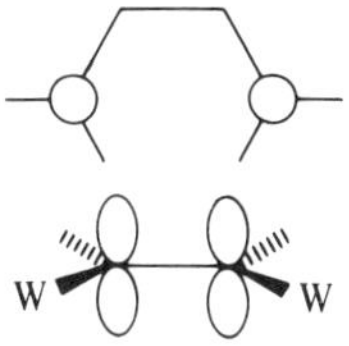

[a] The Translation rotation ED mechanism is shown formally.

a) Both reactants contain the pi bond constrained within a ring such that rotations are prohibited.

b) The reactants have low lying σ^* MO's [21], the presence of which may become responsible for inducing pyramidalization of the appropriate reaction sites.

A comparison of the decay processes involved in the ED, the monorotation, and the pyramidalization mechanisms of a $_2\pi_s + _2\pi_s$ triplet photocycloaddition can be

245

made by reference to Fig. 68. This is a ideogram which depicts the major events involved in the transformation of triplet reactants to singlet products. The ED mechanism involves the most unfavorable decay energy gap factor, while the pyramidalization mechanism involves the best decay energy gap factor since the "hole" which is preserved by this mechanism separates the upper from the lower state by only a small gap, G, which can be smaller than the size of the "hole". Clearly, the ED mechanism will be disfavored relative to the pyramidalization mechanism when ΔE is much larger than G, i.e., decay will be unfavorable due to a large energy gap despite efficient SO coupling. The situation will reverse when $\Delta E \simeq G$. The monorotation mechanism will become important in intermediate cases.

18.7 A Model of Triplet Reactivity

The major theoretical ideas developed in this chapter are the following:

a) A triplet 4N electron photoreaction exhibits a dichotomy insofar as the first part of the reaction, i.e., photochemical barrier crossing, is controlled directly by a one electron *electronic* matrix element of the HO – HO and/or LU – LU type, while the second part, i.e., spin inversion, is controlled by a one electron *SO coupling* matrix element of the HO – LU type. These two different interactions dictate different geometries for the reaction complex. If a rotational mechanism for SO coupling is operative, a triplet 4N electron photoreaction will be partly photoaromatic (photochemical barrier) and partly photoantiaromatic (decay) with photoantiaromaticity imposed by the necessity of spin inversion. This mechanism will result in products having stereochemistry expected from a photoantiaromatic reaction. If the pyramidalization mechanism is operative, the 4N electron photoreaction will be photoaromatic throughout.

b) A triplet 4N + 2 electron photoreaction exhibits no dichotomy. In this case, the photochemical barrier is controlled indirectly by a large HO – LU type one electron *electronic* interaction matrix element. Furthermore, spin inversion is controlled by a one electron *SO coupling* matrix element also of the HO – LU type. Regardless of whether the rotational or pyramidalization SO coupling mechanisms operate, the product stereochemistry will be that expected from a photoantiaromatic reaction.

Fig. 68. Transformation of P.E. surface manifold and resulting decay energy gaps. a) Translational rotational mechanism. b) $2\pi_s + 2\pi_s$ pyramidalization mechanism. $^3\Psi_1$ is the lowest photoantiaromatic triplet and $^1\Psi_0$ is the ground aromatic singlet

c) Reaction polarity modulates the expression of maximum SO coupling potential. In non-polar cases, a large decay energy gap renders unfavorable the mechanisms which afford maximum SO coupling efficiency. In such cases, modest SO coupling efficiency complements an optimally small decay energy gap in the case of pyramidalization mechanism. In polar cases the energy gap factor ceases to be a problem, and mechanisms affording maximum SO coupling can become dominant.

The dependence of stereoselectivity on reaction polarity in the case of triplet photoreactions is spelled out in Scheme 7. The rules embodied in this Scheme are, of course, very approximate but they can hopefully provide the basis for constructive experimentation.

While we exemplified the key concepts involved by reference to photocycloadditions initiated by triplet $\pi\pi^*$ excitation, the basic conclusions remain unaltered in the case of triplet $\pi\pi^*$ involved photocycloadditions initiated by triplet $n\pi^*$ excitation where DA^* $(n\pi^*)$ or D^*A $(n\pi^*)$ is crossed by D^+A^-. Once the reaction system finds itself on the adiabatic surface arising primarily from the D^+A^- diabatic surface the crucial SO coupling matrix element involve HO $-$ LU interaction and the situation resembles the one encountered in triplet $\pi\pi^*$ photocycloadditions. Many triplet $n\pi^*$ photocycloadditions belong to this class. Any polar nonionic triplet photocycloaddition will obey the same rules regardless of the type of excitation.

Finally, it should be pointed out that the SO coupling selection rules for photoaromatic reactions were developed by reference to a model reaction where $^3\Lambda_2$ passes through the "hole" created by the two electron $^1\Lambda_1 - {}^1\Lambda_3$ interaction. Similar conclusions are reached if $^3\Lambda_2$ crosses successively $^1\Lambda_1$ and $^1\Lambda_3$, or, vice versa. In such an event, the reaction system maintains a geometry such that rotation or pyramidalization occurs in the neighborhood of the second crossing.

Most organic chemists have never been exposed to quantum mechanical problems dealing with angular momentum. Thus, this chapter may appear difficult to comprehend. A clear and easy to follow discussion of angular momentum operators and associated matrix elements can be found in the excellent text of McGlynn et al. [24b].

Scheme 7. The stereoselectivity of triplet $\pi + \pi$ photoreactions[a]

Number of electrons	Reactivity spectrum		
	Polar reaction $-$ $-$ $-$ $-$ $-$ $-$ $-$ $-$ $-$ $-$ $-$ $-$ $-$ $-$ Nonpolar reaction		
4N	Predominant s + a	Predominant Diradicaloid	Predominant s + s
4N + 2	Predominant s + s	Predominant Diradicaloid	Predominant s + a

[a] It is assumed that D^+A^- crosses the locally excited diabatic surfaces.

18.8 Spin Orbit Coupling and the Chorochemistry of Triplet Photocycloadditions

The model of triplet reactivity described before represents a very recent development and, hence, remains still untested. However, triplet reactivity trends do exist and seem to conform to the predictions of the model. These are summarized briefly below:

a) Nonpolar triplet $_2\pi_s + _2\pi_s$ stereoselective cycloadditions are known, e.g., entries D3, D7, and F5 of Table 20.

b) Polar triplet cycloadditions leading to unexpected $_2\pi_s + _2\pi_a$ formal adducts are known, e.g., entries B1, F3, and F4 of Table 20.

c) As polarity increases $_4\pi_s + _2\pi_s$ stereoselectivity increases. This is suggested by a comparison of entries B4 and E1 of Table 20.

d) As polarity increases, the ratio of a $4\pi + 2\pi/2\pi + 2\pi$ cycloaddition increases since SO coupling in a $4\pi + 2\pi$ reaction is energetically less expensive than SO coupling in a $2\pi + 2\pi$ reaction. This is suggested by the data referred to in Sects. 8.2 and 8.3.

A more definitive appraisal of the model requires a more secure identification of the excited states involved in the reactions cited above. More importantly, a systematic investigation of the choroselectivity of triplet photocycloadditions, where a gradual change of the electronic properties of the reactants is correlated with a change of choroselectivity, must at long last be undertaken.

18.9 Spin-Orbit Coupling and the Regiochemistry of Triplet Photoaromatic Substitutions

In a previous section we discussed the regioselectivity of singlet photoaromatic substitutions. The regioselectivity of the triplet counterparts can be discussed in a similar manner, now using triplet charge transfer and locally excited diabatic surfaces rather than singlet ones. The differences between singlet and triplet photoaromatic substitution can be understood by reference to the chemical equations shown below:

$$\text{Singlet:} \quad D + {}^1A^* \rightarrow ({}^1D \ldots A^*) \rightarrow [{}^1N'^*] \rightsquigarrow {}^1N_\pi^* \rightarrow {}^1N_\sigma^* \rightarrow \text{Product}$$
$$D + A$$

$$\text{Triplet:} \quad D + {}^3A^* \rightarrow ({}^3D \ldots A^*) \rightarrow {}^3M \rightsquigarrow {}^1N_\pi^* \rightarrow {}^1N_\sigma^* \rightarrow \text{Products}$$
$$D + A$$

In ionic reactions occurring in solution, solvation effects are known to be important. Thus, the ^3M intermediate is reasonably expected to be highly solvated in which case the intermolecular contribution to SO coupling can be very substantial. In addition, many substituted aromatic molecules are known to have short triplet lifetimes, something which indicates a strong intramolecular contribution to SO coupling. Under these circumstances, it seems reasonable to anticipate that the $^3M \rightsquigarrow {}^1N_\pi^*$ conversion will be facile, i.e., this will *not* be the rate determining step in triplet photoaromatic substitutions, at least in most cases. Accordingly, singlet and triplet photoaromatic substitutions may exhibit similar regioselectivity since the regiochemical conditions for optimization of the photochemical barrier are identical for triplet and singlet photoaromatic substitutions. Of course, once again it is assumed that singlet as well as triplet excitation involves HO → LU promotion.

Finally, it should be emphasized that, while in this chapter we were concerned with SO coupling in photoreactions, the same considerations apply to thermal reactions where the ground singlet P.E. surface is crossed by the lowest triplet P.E. surface. At the ZIDMOO approximation level, surface "touchings" rather than crossings materialize and the reader is warned of this shortcoming of our theoretical approach. Typical thermal reactions of singlet molecules which may yield triplet intermediates are geometric isomerizations and ionic antiaromatic reactions.

19. Photophysical Processes

A photoexcited molecule can interact with a ground state molecule to yield a product. However, this interaction may have other consequences aside from reaction. These other alternatives have already been implicit in the P.E. surfaces we have constructed. In the space below, a more explicit discussion is given of the important photochemical processes which result into some form of energy transfer not accompanied by reaction.

I. Quenching

An excited molecule can be quenched by a ground state molecule and this process can be easily understood by reference to the P.E. surfaces constructed before for the purpose of discussing photoreactions, e.g., Fig. 27. The equation for quenching, shown below, is nothing else but part of the equations for P2' or P2'' nonionic photoreactions:

$$
\begin{array}{c}
\mathbf{D + A} \\
\uparrow \\
\mathbf{D + A^*} \rightarrow (\mathbf{D \ldots A^*}) \longrightarrow \mathbf{M} \longrightarrow \text{reaction} \\
\downarrow \\
\mathbf{D + A} + h\nu
\end{array}
$$

The dependence of the rate of fluorescence or phosphorescence quenching of A^* by a ground state molecule D on polarity can be understood by reference to the simplified kinetic scheme shown below, and our previous analysis of photoreactivity (see Chap. 4).

$$
\mathbf{D + A^*} \underset{k_{-1}}{\overset{k_1}{\rightleftharpoons}} \text{Excited Complex (e.g., M)} \begin{array}{c} \overset{k_2}{\longrightarrow} \text{Products} \\ \underset{k_3}{\longrightarrow} \mathbf{D + A} \end{array}
$$

As polarity increases, k_1 and k_3 increase while k_{-1} decreases. On the other hand, k_2 may remain relatively insensitive (photoaromatic quenching) or also increase (photoantiaromatic quenching). Accordingly, we predict that the rate of quenching will increase as polarity increases. Table 45 contains data in support of this prediction[1].

The dependence of the rate of quenching on the energy of the photoexcited molecule can be simply predicted on the basis of considerations outlined in Chap. 3. Specifically, in comparing the rate of quenching of A^* and A'^* by D and assuming that the diabatic surface DA'^* is translated upwards in energy relative to DA^* while the D^+A^- and $D^+A'^-$ diabatic surfaces remain nearly superimposable, we predict that the interaction of D and A'^* will involve a lower photochemical barrier and that the rate of quenching will be greater. Experimental results which seem to be in accordance with these predictions can be found in the literature [135].

An emitting excited complex, say M, can be quenched by a ground state molecule Q. Recalling that M is a resonance hybrid of the type $D^+A^- \longleftrightarrow DA^* \longleftrightarrow D^*A \longleftrightarrow D^-A^+$, we can predict the structure of the trimeric excited species $[DAQ]^*$ which is involved in the quenching. Thus, when Q is a good donor, the most important configuration mixing will be that of QD^+A^- and Q^+DA^-. The corresponding interaction matrix element is the $HO^Q - HO^D$ type. Accordingly, it is predicted that the more stable isomer will be $[QDA]^*$ rather than $[DAQ]^*$. On the other hand, if Q is a good acceptor, the reverse will be true.

The quenching of the fluorescence or phosphorescence of an excited complex can be discussed along similar lines as that for an excited molecule. The conclusions remain the same, i.e., quenching will become more efficient as the donor or acceptor ability of the quencher increases.

II. Energy Transfer

An excited molecule can transfer its energy to a ground state molecule, and the mechanism involved in each case can be easily understood by reference to the P.E. surfaces shown in Fig. 69 which depicts singlet-singlet or triplet-triplet energy transfer from D to A. The characteristic chemical equations are as follows:

$$D^* + A \longrightarrow (D^* \ldots A) \longrightarrow M'^* \rightsquigarrow D \ldots A^* \longrightarrow D + A^*$$
$$M \longrightarrow \text{etc.}$$

A complete understanding of the energy transfer mechanism is not possible because the partition of M' cannot be predicted reliably. An analysis for singlet-triplet energy transfer can be given along the same lines assuming that spin orbit coupling will allow the system to make a transition from singlet to triplet surfaces.

---

1   It should be noted that the rate of quenching of fluorescence can vary all the way to the diffusion control limit. This limit is anticipated on the basis of the P.E. surfaces constructed by the LCFC method in P2' and P2'' photoreactions where the D^+A^- intersects DA^* early on the reaction coordinate and across its nearly flat segment.

Table 45. Quenching constants of anthracene and pyrene ($\pi\pi^*$) fluorescence by acceptors *(1)*

Quencher	A (eV)	Pyrene [(kcal/mole)]	Anthracene [(kcal/mole)]
(phthalic anhydride structure)	0.15	385	122
(maleic anhydride structure)	0.57	770	35[a]
Cl$_4$ (tetrachloro indanedione structure)	0.58	500	227
(benzoquinone structure)	0.60	833	185
1.3.5 (NO$_2$)$_3$ (trinitrobenzene structure)	0.70	909	278
Cl, Cl (dichlorobenzoquinone structure)	1.15	1,110	370
Cl$_4$ (tetrachlorobenzoquinone structure)	1.37	–	1,100

[a] Small, due to Diels Alder reaction.

(1) Nakayama, A., Akamatu, H.: Bull. Chem. Soc. Japan. *41*, 1961 (1968).

III. Intersystem Crossing By Complexation

A singlet molecule, $^1A^*$, can be converted to a triplet, $^3A^*$, radiatively or by inter-action with another molecule, D. This latter situation is illustrated in Fig. 69. The characteristic chemical equations are as follows:

$$D + {}^1A^* \longrightarrow (D \ldots {}^1A^*) \xrightarrow[\text{Crossing}]{\text{Intersystem}} {}^3M'^* \rightsquigarrow (D \ldots {}^3A^*) \longrightarrow D + {}^3A^*$$

$$\searrow {}^3M$$

$$\downarrow$$

Etc.

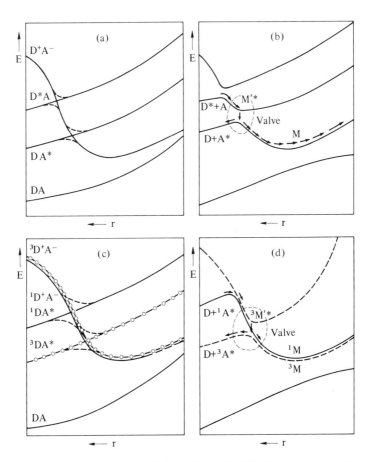

Fig. 69. a) Diabatic surfaces and their interactions (indicated by dashed lines) for energy transfer between two molecules. b) The corresponding adiabatic surfaces and valve. c) Diabatic surfaces and their interaction (indicated by dashed lines) for intersystem crossing due to complex formation. d) The corresponding adiabatic surfaces and valve

253

At this point, attention should be called to the fact that, if D*A lies *underneath* DA*, singlet-singlet energy transfer will occur, while, if ^{3}D*A lies *above* ^{3}DA*, singlet-triplet energy transfer will occur. Finally, if both of the aforementioned situations materialize, intersystem crossing, singlet-singlet and singlet-triplet energy transfer will all compete. It is, then, obvious that, for efficient energy transfer or intersystem crossing, the two key diabatic surfaces should not sandwich any other diabatic surfaces which may give rise to competing processes.

Two typical sequences of photophysical processes initiated by singlet excitation of **A** and dictated by the relative energies of the locally excited diabatic surfaces are shown in Scheme 8. Assuming that all the four possible photochemical processes are P2' or P2'', it is most likely that, in Case I, singlet-singlet energy transfer will predominate, while in Case II, intersystem crossing will be most efficient. This arises because, for every transition from a higher to a lower surface, the photoexcited composite system has to go through a "valve" which diverts it either to photoexcited components or a photoexcited intermediate which may decay or react further.

Scheme 8. Possible energy transfer processes following excitation of molecule A

I. ^{1}DA*

      ^{1}D*A   Singlet-singlet energy transfer

      ^{3}DA*   Singlet-triplet energy transfer

      ^{3}D*A*   Triplet-triplet energy transfer

II. ^{1}DA*

      ^{3}DA*   Intersystem crossing

      ^{1}D*A   Singlet-triplet energy transfer

      ^{3}D*A   Intersystem crossing

254

20. The Importance of Low Lying Nonvalence Orbitals

In this book, we have assumed uniformly that the LU's of unsaturated molecules are of the valence (V) type. However, it is well known that in certain molecules the lowest excited state has Rydberg character. Accordingly, one should inquire as to how the existence of such states alters the conclusions reached by considering only valence orbitals. The dynamic LCFC method is an ideal framework for discussing differences in chemical reactivity due to participation of V or Ry orbitals and an abbreviated analysis is presented below.

We shall illustrate our approach by reference to a chemical reaction adequately described by a minimal basis set of DA, D^+A^-, and DA^* diabatic surfaces, where D^+A^- and DA^* can involve promotion of an electron to a V or Ry LU of the acceptor. Focusing first on the thermal reaction, we distinguish the following two cases:

a) D^+A^- (V) has lower energy than D^+A^- (Ry). This case has been discussed in the previous chapters and may constitute the most frequently encountered situation in thermal organic reactions. Here, the $HO^D - LU^A$ (V) interaction dictates primarily the properties of the ground diabatic surface.

b) D^+A^- (V) has higher energy than D^+A^- (Ry). In this case, the $HO^D - LU^A$ (Ry) interaction dictates primarily the properties of the ground adiabatic surface. This situation can become qualitatively similar to the previous one under any of following circumstances:

1. LU^A (V) has the same symmetry as LU^A (Ry).
2. The $HO^D - LU^A$ (Ry) interaction is weak because the corresponding overlap integral is small. In this case, although the $DA^- - D^+A^-$ (Ry) energy gap is smaller than the $DA - D^+A^-$ (V) energy gap, the properties of the ground adiabatic surface are controlled by the $HO^D - LU^A$ (V) interaction.

We now enter the discussion of photochemical reactions. Here, we should be prepared to consider the most general situation where the energy of a LU depends upon the occupancy of the molecule. Thus, whenever LU(V) and LU (Ry) are closely spaced, it is conceivable that DA^* (V) has lower energy than DA^* (Ry) while D^+A^- (Ry) has lower energy than D^+A^- (V), and vice versa. We distinguish the following cases.

a) D^+A^- (V) has lower energy than D^+A^- (Ry) and DA^* (V) has lower energy than DA^* (Ry). These cases have been discussed in the previous chapters and may constitute the majority in organic photochemistry. Here, the $HO^D - HO^A$ interaction dictates the properties of the lowest excited adiabatic surface.

b) D^+A^- (V) has lower energy than D^+A^- (Ry) but DA^* (V) has higher energy than DA^* (Ry) or D^+A^- (V) has higher energy than D^+A^- (Ry) but DA^* (V) has lower energy than DA^* (Ry). In both cases, the features of the lowest excited adiabatic surface will be dictated by the two electron mixing of valence and Rydberg adiabatic surfaces and specific covalent bonding will be absent.

c) D^+A^- (V) has higher energy than D^+A^- (Ry) and DA^* (V) has higher energy than DA^* (Ry). This is an interesting situation which will resemble situation a) because the D^+A^- (Ry) $- DA^*$ (Ry) interaction matrix element, responsible for the mixing of the corresponding diabatic surfaces to produce the lowest adiabatic excited surface is of the $HO^D - HO^A$ type.

From the above discussion, it is clear that the participation of Ry orbitals can be felt only in case b). However, even in this case, valence orbital interaction may dictate the properties of the lowest excited adiabatic surface. For example, although the energy gap favors stronger DA^* (V) $- D^+A^-$ (Ry) or DA^* (Ry) $- D^+A^-$ (V) interaction, the DA^* (V) $- D^+A^-$ (V) or DA^* (Ry) $- D^+A^-$ (Ry) interaction may predominate due to a larger $HO^D - HO^A$ one electron interaction matrix element. Finally, nonvalence orbitals may not have any chemical significance in photoreactions where the various reactivity aspects are determined by a decay process controlled by the interaction of no bond and diexcited surfaces.

In short, Rydberg excitation may lead to the same chemical results as valence excitation depending upon the electronic properties of the reactants. On the other hand, situations are predicted to exist where the two types of excitation may lead to different reactions. While the similarities and differences discussed above are connected with the shapes of the crucial adiabatic surfaces, differences and similarities may also be due to different or similar conformational properties of Rydberg and valence excited states.

In the space above, we have set up rather than solve the problem of nonvalence orbital participation in chemistry. A solution is at the present time impossible because little or no information exists regarding the electronic nature of radical anions and lowest excited states of organic molecules [136]. Progress in this area is crucial to a fuller understanding of chemical reactions and it is hoped that the discussion presented here will stimulate spectroscopists and theoreticians to pursue these problems with greater vigor. Some interesting ideas regarding the problems alluded to before have already been expressed [137].

21. Divertissements

21.1 Thermal Antiaromatic Pericyclic Reactions

Consider the case of $2\pi + 2\pi$ dimerization of $CH_2 = O$ as the simplest model reaction system for illustrating the key principles. It is predicted that the reaction will be HT regioselective if it occurs in a $_2\pi_s + _2\pi_a$ manner controlled by HO − LU interpacket interactions. On the other hand, if the reaction occurs in a $_2\pi_s + _2\pi_s$ manner, HT regioselectivity will be imposed if HO − LU interpacket interactions control the barrier height, while HH regioselectivity will be imposed if LU − LU and HO − HO intrapacket interactions are the most important factor. Now, the interesting aspect of this problem is the fact that a HT $_2\pi_s + _2\pi_s$ union will involve pericyclic bonding due to the additive effect of the HO − LU interactions of the two molecules imposed by the nature of the $\langle \Psi_1 | \hat{P} | \Psi_2^+ \rangle$ type matrix element. A pictorial illustration is given below. Note that the bonding interaction of one pair of uniting centers imposed by one HO − LU interaction outweighs the antibonding interaction between the same pair

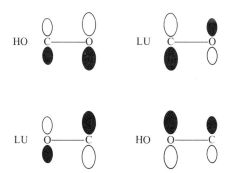

imposed by the second HO − LU interaction. Accordingly, we predict that, if HO − LU interactions dominate, one should be able to observe a $_2\pi_s + _2\pi_s$ stereoselective and HT regioselective $2\pi + 2\pi$ cyclodimerization of a heterodensic molecule.

21.2 Induced Thermal Pericyclic Reactions

In certain reactions which are expected to proceed in a quasipericyclic s + s manner, an orbital other than those apparently involved in the electronic reorganization may enter into play and enforce a geometry of approach which is characterized by s + s pericyclic bonding. This "extra orbital" trick is illustrated by the reaction shown below. Here, we need to consider only the three configurations DA, D^+A^-, and $D^+A'^-$, where DA is the normal no bond configuration, D^+A^- involves electron trans-

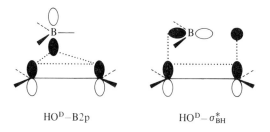

D   . A

fer from HO^D to the vacant B2p AO, and $D^+A'^-$ involves electron transfer from HO^D to σ^*_{BH}. The thermal reaction will proceed in a way which maximizes interpacket interactions, i.e., the $HO^D - B2p$ and $HO^D - \sigma^*_{BH}$ interactions which are maximized for the geometry shown above. Both types of interactions give rise to pericyclic bonding as the interaction matrix elements sketched below suggest. Hence, the reaction is expected to be $_2\pi_s + _2\sigma_s$ stereospecific.

$HO^D - B2p$ $HO^D - \sigma^*_{BH}$

In the hypothetical case where the vacant B2p AO is not present and, hence, the $HO^D - B2p$ interaction is nonexistent, a T-like geometry of approach would be favored over the $_2\pi_s + _2\sigma_s$ one as discussed in Chap. 9. In short, the presence of the "inductor" orbital B2p enforces a geometry of approach which would not be preferred in its absence.

21.3 Half-Aromatic Reaction Paths

In this work, we have made a distinction between aromatic and antiaromatic LM and NLM paths. LM aromaticity and LM antiaromaticity are imposed by the symmetry properties of the MO's of the reactants as exemplified below.

LU A——   ——A LU │ LU A——   ——S LU

HO S—++   ++—S HO │ HO S—++   ++—A HO

LM Antiaromatic path   │   LM Aromatic path

Simply stated, an LM path is favored because interpacket interactions weaken the existing bonds of the reactants and create new bonds between them along the entire reaction coordinate. This is not possible in an LM antiaromatic path. However, an intermediate situation may arise when the symmetry properties of the reactant MO's are as follows:

LU's——►   S——   ——A (S)   A——   ——A (S)

HO's——►   S++   ++—S (A)   A++   ++—S (A)

These reactions can now proceed via an LM path which would involve some reactant bond weakening and some interbond formation. Molecules or fragments which involve HO and LU of identical symmetry have been recently identified [21], e.g., a sigma C–C bond, and their reactions should have characteristics intermediate to those of pericyclic and quasipericyclic reactions. If the dominant HO – LU interaction, due to the $DA - D^+A^-$ mixing, favors interbonding, these reactions should resemble thermal LM aromatic reactions.

21.4 Ambient Reactivity

Consider the two nucleophilic substitution reactions shown below:

$$R-X + NO_2^- Ag^+ \rightarrow ONO-R + O_2N-R + AgX$$

$$\text{1a} \qquad \text{2a}$$

$$R'-X + NO_2^- Ag^+ \rightarrow ONO-R' + O_2N-R' + AgX$$

$$\text{1b} \qquad \text{2b}$$

$$R = O_2NC_6H_4CH_2^-$$
$$R' = CH_3OC_6H_4CH_2^-$$

The first reaction can be thought of as S_N2 – like and the second as nucleophilic attack on tight ion pair. The transition state of the second, more polar, reaction will occur at longer intermolecular distance. The HO^D-LU^A interaction matrix element

259

will be appreciable in the first case (tight transition state) and will tend to dictate preferential formation of **2a** because nitrogen is the site of highest HO^D electron density. By contrast, the same matrix element will be small due to weak spatial orbital overlap in the second case (loose transition state). Hence, electrostatic effects will tend to dictate preferential formation of **1b** because oxygen is the site of highest negative charge. An interesting corollary is the following: a shift from "orbital" to "charge" control is expected as reaction becomes increasingly faster. In this connection, it has been observed [138] that when $Ag^+NO_2^-$ reacts with an S_N2 prone alkyl halide the HO electron density of the nucleophile determines the major product; when an S_N1 prone alkyl halide is involved, coulombic effects take over.

The agreement of theory and experiment is encouraging but much remain to be done before a complete picture emerges. For example, the thorough investigations of Kornblum et al suggest that solvent [139a], reaction phase [139b], and counter-ion [138] play key roles in determining ambient nucleophilicity.

21.5 The Stereoselectivity of Photocycloadditions: In Search of Antiaromatic Intermediates

In 4N electron cycloadditions, the barrier of the LM thermal reaction is created from the two electron interaction of the reactant ground and diexcited surfaces. The qualitative nature of our approach and, especially, the ZIDMOO approximation in the evaluation of the interaction matrix elements does not allow any simple predictions regarding the details of the adiabatic surfaces. For example, one may envision a shallow dip housing an Antiaromatic Intermediate (AI) on top of the thermal $_2\pi_s + _2\pi_s$ cycloaddition, which could undergo bond rotation and become responsible for non-stereospecificity. Is there any reliable experimental procedure for detecting the presence of such an intermediate?

A nonstereoselective 4N electron thermal cycloaddition can, in principle, be compatible with a $_2\pi_s + _2\pi_s$, TB, $_2\pi_s + _2\pi_a$, or mixed mechanism simply because the least motion path is placed at an "electronic disadvantage" and a "steric advantage" relative to the non least motion path. Thus, stereochemical studies are not expected to be illuminating. By contrast, in a 4N electron photochemical cycloaddition, the LM path is favored both electronically and sterically. Accordingly, such reactions should by s + s stereoselective. However, this simple picture needs modification if we admit the presence of an AI intermediate. In such an event, the mechanism of the reaction becomes as follows:

$$D + A^* \to (D...A^*) \to M \to \Xi'^* \to AI \overset{\displaystyle \to D + A}{\underset{\displaystyle \to \text{Products}}{<}}$$

Partial or complete loss of the stereochemical integrity of the reactants can now be taken to signify the existence of an AI intermediate. The existing chemical data suggests that some 4N electron photocycloadditions are stereospecific or highly stereo-selective in an s + s sense while others are not (see Table 20). It is possible that a better understanding of thermal 4N electron cycloadditions will emerge from studies of the corresponding photochemical reactions.

The same concepts are applicable to 4N electron multicentric reactions where the steric requirements of the LM and NLM paths are significantly different. This excludes electrocyclizations from such considerations. Also, it should be emphasized that the same analysis is valid if the shallow dip on the thermal barrier top is replaced by its operational equivalent, namely, a flat barrier top.

22. A Contrast of "Accepted" Concepts of Organic Reactivity and the Present Work

The statement, "entities are not to be multiplied without necessity", has often been ascribed to William of Ockham and has become known as Ockham's razor. It is a statement about overcompleteness and its applicability to science has been recognized. Accordingly, in submitting this work to public scrutiny, we have to defend the position that it does not constitute a restatement of accepted ideas. One operational test of novelty is the disparity between the overview developed on the basis of the allegedly novel contribution and the overview which is accepted by the majority. Hence, what follows is a presentation of examples of the differences between our approach and what I perceive to be the "accepted viewpoint". A complete list cannot be presented due to space limitations.

I. A diradical, as understood by a theoretician, is an entity having two nonbonding electrons. On the other hand, a diradicaloid may be defined as an entity having two weakly bonding electrons. Finally, organic chemists frequently use yet another term, herein denoted as "diradical", which is left undefined and could be given the significance of the term diradical or the term diradicaloid. In plain language, the term "diradical" is a catch-all term. Its ambiguity is further accentuated when we realize that antiaromatic *as well as* nonaromatic paths may involve diradicals and diradicaloids. For example, consider the case of a nonionic $_2\pi_s + _2\pi_s$ cycloadditions. In the vicinity of the pseudocrossing of the DA diabatic surface and the Λ_3 boundary, the reaction complex assumes the properties of a diradical, while prior or after the pseudocrossing a diradicaloid may be formed. As a second example, consider the case of a TB cycloaddition. Here, the initial TB intermediate is a diradicaloid which can be transformed to another diradicaloid via a diradical. The delocalized equivalent of this picture has been discussed by Dewar [140].

The reader can now appreciate my disdain for the term "diradical" which actually means everything and nothing. Specifically, by stating that a reaction proceeds via a "diradical" one does not differentiate whether it proceeds via an antiaromatic or a nonaromatic path. Most authors have used the "diradical" construct in order to circumvent Woodward-Hoffmann forbiddenness and what they actually implied is that the reaction proceeds via a nonaromatic path. I suggest that these interpretations are incorrect in the case of many nonionic and most ionic thermal 4N pi electron cycloadditions.

The best example of the pressure exerted by the Woodward-Hoffmann rules to discourage consideration of antiaromatic paths as viable mechanistic possibilities comes from the area of ionic $2\pi + 2\pi$ cycloadditions. Thus, most chemists have adopted the viewpoint that thermal ionic $2\pi + 2\pi$ cycloadditions occur via dipolar intermediates devoid of pericyclic bonding. By contrast, we suggest that these reactions can be thought of as true pericyclic reactions, for all practical purposes. Furthermore, our analysis suggests that the dipolar intermediate is nothing else but an exciplex which descended to the ground surface and that thermal ionic $2\pi + 2\pi$ cycloadditions can be thought of, in a sense, as "thermal photochemical reactions".

Similar objections can be raised for the interpretation of photochemical reactions via the "diradical" construct. On the basis of the present work, it is now possible to provide a more specific answer to whether a reaction is nonaromatic or antiaromatic since these two paths have often different chorochemical consequences.

II. A stereospecific or highly stereoselective thermal reaction which violates the Woodward-Hoffmann rules has been assumed to proceed via a pathway involving intermediates devoid of pericyclic bonding or via photochemical activation. By contrast, we have argued the following:

a) Hückel antiaromatic paths can be preferred over sterically hindered Möbius aromatic as well as nonaromatic paths in thermal EE and OO multicentric reactions.

b) Aromatic, nonaromatic and antiaromatic paths are closely spaced in energy in thermal EO multicentric reactions. Accordingly, these are expected to be least motion reactions.

c) All Woodward-Hoffmann "forbidden" reactions can be made pericyclic (or, effectively pericyclic) by placing appropriate substituents on the reactants. The thermal reactions of Table 46 are predicted to be pericyclic, or, effectively pericyclic and, as a result, highly stereoselective. These conclusions are opposite to the ones which one would arrive at by application of the Woodward-Hoffmann rules.

III. Most chemists have associated a single-hump ground P.E. surface with an aromatic reaction and a double-hump ground P.E. surface with a stepwise nonaromatic reaction. By contrast, we predict secondary minima in all aromatic, nonaromatic and antiaromatic *ionic* reactions.

IV. The rate of a chemical reaction is a topic of paramount importance to every practicing chemist. Accordingly, many attempts have been made to correlate reaction rates with a given property of the various substrates. These attempts have met with partial success to the extent that a certain trend can be discerned, in most cases, but exceptions always abound. This is understandable because reaction polarities as well as interaction matrix elements should be considered in making a prediction about relative rates. Thus, we should distinguish between two different types of comparisons of reaction rates:

a) Comparisons where the difference in the polarity, ΔP, of the two systems which are being compared is zero or very small. In this case, the size of the matrix element determines relative reactivity. An example of a comparison where $\Delta P \simeq 0$ is any comparison of two chorochemical (not topochemical) paths of a given reaction.

Table 46. Woodward-Hoffmann "Forbidden" thermal pericyclic or effectively pericyclic reactions

Number of electrons	Reaction	Path	Condition for pericyclicity
4N	$\pi + \pi$ Cycloadditions	s + s	Ionic reaction
4N + 2	$\pi + \pi$ Cycloadditions	s + a	Ionic reaction
4N	$\pi + \sigma$ Addition	s + s	Ionic reaction
4N	$\pi + \pi$ Electrocyclization	Disrotary	Ionic reaction
4N + 2	$\pi + \pi$ Electrocyclization	Conrotary	Ionic reaction
4N − 1	$\pi + \pi$ Cycloaddition	s + s	−
4N + 1	$\pi + \pi$ Cycloaddition	s + a	−
4N − 1	$\pi + \sigma$ Addition	s + s	−
4N − 1	$\pi + \pi$ Electrocyclization	Disrotatory	−
4N + 1	$\pi + \pi$ Electrocyclization	Conrotatory	−
4N	Sigmatropic shift	SR	Low lying $D^{+*}A^-$ or $A^{-*}D^+$
4N + 2	Sigmatropic shift	SI	Low lying $D^{+*}A^-$ or $A^{-*}D^+$
4N	Ionic rearrangement	SR	Lowest energy open shell configuration
4N	Ionic rearrangement	SI	Lowest energy open shell configuration
4N + 1	Radical rearrangement	SR	Electrophilic radical
4N − 1	Radical rearrangement	SI	Nucleophilic radical

b) Comparisons, where ΔP is very large. In this case, polarity alone determines relative reactivity, i.e., as $I_D - A_A$ decreases reaction rate increases.

On the basis of the above considerations, we can disavow a "rule of thumb" often employed by organic chemists. Specifically, the Woodward-Hoffmann concept of "forbiddenness" and "allowedness" is valid only for comparisons where ΔP is small, e.g., a Woodward-Hoffmann "allowed" $_\pi 4_s + _\pi 2_s$ cycloaddition will be faster than a "forbidden" $_\pi 2_s + _\pi 2_s$ cycloaddition, *if the two reactions have comparable $I_D - A_A$.* On the other hand, the same concept is invalid for comparisons where ΔP is large, e.g., a "forbidden" $_\pi 2_s + _\pi 2_s$ cycloaddition where $I_D - A_A$ is small can be faster than an "allowed" $_\pi 4_s + _\pi 2_s$ cycloaddition where $I_D - A_A$ is large.

V. Houk [15] has expressed certain viewpoints regarding the effect of symmetrical versus unsymmetrical substitution on the mechanism of cycloaddition reactions. For example, he proposed that nonaromatic paths can be preferred over aromatic paths in highly polar 4N + 2 pi electron cycloadditions, if at least one reactant is unsymmetrical, leading to formation of dipolar intermediates. However, the examples cited are not sufficiently unsymmetrical to give rise to such preference, e.g., $Ph_2C=CH-CH=CH_2$ plus $(CN)_2C=C(CN)_2$. By contrast, we predict that pericyclic N^* intermediates will be formed in ionic reactions even when both reactants are symmetrical. However, loss of stereochemistry will be more likely if unsymmetrical substitution makes one bond stronger than the other.

VI. Quantitative rate comparisons should be limited to reactions of the same type. For example, in type A and A' reactions, the experimental rate constant is a function of a bimolecular rate constant, while in type B reactions, it is a function of an equilibrium constant and a unimolecular rate constant (see Fig. 32). Thus, a quantitative comparison of a type A and a type B reaction is ill advised since different parameters are involved. On the other hand, qualitative comparisons are possible since in most cases these parameters vary in the same direction as chorochemistry or polarity is varied.

VII. The various attempts to interpret the regioselectivity of multicentric reactions, successful as they have been, were based on the original work of Fukui [11], according to which the preferred regiochemistry of a reaction is the one which involved union of the two sites of highest frontier electron density. This is valid for bicentric reactions. However, in multicentric reactions, one has to consider the form of the appropriate algebraic expression. This leads to the conclusion that regioselectivity will be greatest in Woodward-Hoffmann "forbidden" reactions.

VIII. The HH versus HT regiochemistry of thermal and photochemical cycloadditions has been interpreted by postulating the formation of a discrete "diradical" or dipolar intermediate which is preferentially formed in a manner which optimizes the stabilizing effect of the substituents. While this hypothesis has been operationally useful, it must now be rejected as unrealistic for the following reasons:

a) The regiochemistry of thermal cycloadditions which proceed without formation of discrete intermediates, e.g., Diels-Alder reactions, can be rationalized theoretically.

b) The regiochemistry of many cycloadditions is predicted and found to be opposite to the one expected on the basis of the hypothesis of the "diradical" intermediate formation.

IX. While Hammond's postulate [141] implies that the selectivity of a reaction decreases as reaction rate increases, our conclusions are different. Specifically, the dependence of selectivity on reactivity depends on whether a thermal reaction is non-ionic or ionic. In the former case, selectivity increases as reactivity increases, while in the latter case, the reverse situation obtains.

X. Klopman [12] has advanced a theoretical treatment, using the static model, which differentiates between charge-controlled and frontier-controlled EE reactions. This author concluded that high reactivity can be accompanied by either of the two types of controlling influences. By contrast, our approach predicts that slow EE reactions, involving tight transition states, will tend to be energy gap or matrix element frontier controlled and fast ionic EE reactions, involving loose transition states, where orbital overlap effects are deemphasized, will tend to be subject to coulombic effects.

XI. Photochemists have attempted to interpret the stereochemistry of photochemical multicentric reactions by reference to the Woodward-Hoffmann rules. On the other hand, we have identified classes of photoreactions which are expected to occur stereochemically in a manner different from that predicted by Woodward and Hoffmann. These are dual channel singlet photocycloadditions initiated by $\pi\pi^*$ or $n\pi^*$ photoexcitation and a yet unknown number of triplet photocycloadditions.

XII. Various workers have utilized the reaction of two radicals Y· to form a covalently bonded molecule Y_2 as the prototype for drawing inferences regarding the reactivity differences between singlet and triplet photoreactions [142, 143]. According to the P.E. surfaces of Fig. 57a, the lowest excited triplet adiabatic surface minimum occurs at infinite intermolecular distance while the lowest excited singlet adiabatic surface displays a minimum at tight geometry. Taking note of the different location of the triplet and excited singlet minima, it was proposed that singlet and triplet excited Woodward-Hoffmann "allowed" pericyclic reactions differ because the geometry of the corresponding triplet intermediate is loose giving rise to stereorandomization while the corresponding singlet intermediate is tight and capable of stereospecific product formation. An example of the application of this proposal principle taken from the work of Michl is given below.

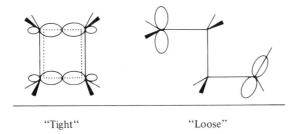

"Tight"   "Loose"

Focusing attention on the particular example shown above, we can immediately pinpoint that the analogy between the Y_2 and C_4H_8 decompositions is not valid. This arises because in the former case the triplet no bond diabatic state cannot mix in a one-electron sense with any triplet charge transfer diabatic state. By contrast, in the case of C_4H_8 decomposition, one electron mixing of $^3\Psi_2^+$, $^3\Psi_2^-$, $^3\Psi_3^+$ and $^3\Psi_3^-$ diabatic surfaces is possible, giving riste to a barrier followed by a tight triplet excited intermediate. In short, the analysis of the stereochemistry of pericyclic reactions based on the consideration of the $Y_2 \rightarrow 2Y$ model system is not valid insofar as it neglects what may well be the most important factor, namely, the one electron interaction of triplet locally excited and charge transfer diabatic surfaces.

XIII. In a recent article, Michl et al. [142] posed the question: "What is the relation of the interaction and correlation diagrams contructed using MO's and/or states of the fragments representing partners in cycloaddition reactions to the "supermolecule" correlation diagram approach?" I feel that this work provides a rather clear cut answer to this important question.

In addition, Michl et al. [144] pointed out, in reference to our early work using the static model approximation, that the very important diexcited configurations are frequently missing in discussions of the fragment model. However, it should now be clear that this omission does not affect the conclusions of our original qualitative analysis because the relative energetics of the crucial transition states can be well understood even when diexcited configurations are neglected. On the other hand, inclusion of these configurations is necessary for a satisfactory understanding of certain decay processes. Nonetheless, in most cases minimization of a barrier height and maximization of the efficiency of an ensuing decay process are obtained for the same chorochemistry. This constitutes the basis of the success of our earlier treatment based on the static model approximation and neglect of diexcitation.

In the same work, Michl et al. [144] suggest that omitting the diexcited surfaces leads to incorrect state correlation diagrams apparently alluding to Fig. 2 of ref. [43]. Needless to say, this statement taken by itself is absolutely right. However, our primary concern at that point was to show how an intermediate may arise from a crossing of the DA and D^+A^- diabatic surfaces at long intermolecular distance. For this purpose, omission of the diexcited surfaces is permissible. Confusion may have arisen due to the fact that the repulsive parts of the charge transfer diabatic surfaces were not shown and no distinction was made between nonionic and ionic reactions.

Turning to self criticism, it can be noted that the LCFC formalism is essentially a Molecular Orbital-Valence Bond-Configuration Interaction type of formalism. Reduction to a simple qualitative theory can be accomplished in one of the following ways:

a) Interfragment differential MO overlap is included but CI is not performed. Such an approach can reveal only the importance of exchange effects.

b) Interfragment differential MO overlap is neglected but CI is performed. Such an approach projects the importance of CI effects.

Approach (a) is indefensible because it cannot account for many well known reactivity trends. Accordingly, we have chosen approach (b) for an initial attack on reactivity problems.

How will rejection of the ZIDMOO approximation, i.e., inclusion of exchange effects, improve the understanding of reactivity already obtained on the basis of the approximate theory expounded in this work? Some deficiencies of the LCFC-ZIDMOO method, which do not affect qualitative conclusions, are discussed below.

a) At the level of the LCFC-ZIDMOO approximation, avoided crossings may become pseudocrossings. These cases are rather infrequent since most crossing diabatic surfaces arise from configurations which differ in the occupancy of a single spin MO and their interaction can be adequately described by the LCFC-ZIDMOO method.

b) The LCFC-ZIDMOO method may introduce unreasonably deep energy wells at long intermolecular distances in all diabatic surfaces where both components are closed shell or one open and the other closed shell. This can arise because interfragment nucleus-electron attraction may dominate nucleus-nucleus and electron-electron repulsion at such distance. Inclusion of exchange terms brings into fore overlap repulsion which can counteract this effect. The correct result is that weak molecular complexes may or may not be formed depending on the case at hand. Cognizant of this limitation of the LCFC-ZIDMOO approach, we have omitted weak complex minima in drawing certain diabatic surfaces but noted their possible existence in the chemical equations (in parenthesis). The important thing is that these weak complexes are known from experimental studies to arise at long intermolecular distances much before the barrier peaks crucial to our arguments.

c) The LCFC-ZIDMOO method *may* produce an incorrect shape for a diabatic surface of the no bond or local excitation type where both components are open shell. In such a case, exchange stabilization *may* dominate exchange destabilization (assuming that the two singly occupied MO's have congruent symmetry) and a minimum may arise. For simplicity, we have drawn diabatic surfaces of this type as repulsive assuming that CI correction ultimately leads to the correct shape of the corresponding adiabatic surface. We find this to be generally true for most chemical problems of interest. This limitation does not affect qualitative arguments and an example will be given below.

d) The LCFC-ZIDMOO method may produce similar shapes to those produced by the rigorous LCFC method with differences in the exact slopes or positioning of minima.

The results of a comparative study are summarized in Table 47. These are restricted to three important types of diabatic surfaces which play a pivotal role in all reactions. Items 4 and 6 illustrate major modifications which, however, do not affect any of the qualitative conclusions. For example, the LCFC-ZIDMOO method predicts that covalent bond strength depends primarily on the $DA - D^+A^-$ interaction, a $HO^D - HO^A$ interaction. On the other hand, the rigorous LCFC method predicts that the same quantity *may* depend partly on the exchange stabilization of the DA diabatic surface which, in turn, depends also on $HO^D - HO^A$ interaction. Of course, part of covalent bonding is also due to $DA - D^+A^-$ interaction according to the latter method. However, even in a quantitative sense, the problem can be treated at the LCFC-ZIDMOO level by appropriate parametrization of the $HO^D - HO^A$ matrix element, i.e., if the ZIDMOO approximation "turns off" any exchange stabilization of the DA configuration, this can be counteracted by appropriate parametrization of β.

By contrast, items 1 and 2 illustrate minor modifications which nonetheless reveal new nuances. For example, these two items taken together suggest, *inter alia,* the following possibilities:

1. Reaction ionicity depends on whether an aromatic or antiaromatic path is followed. This implies that reactions will be found where the assumption of a common intermediate partitioning to Woodward-Hoffmann "allowed" and "forbidden" products will prove incorrect.

2. Dual channel photochemical mechanisms are feasible because the minimum housing the O intermediate (see Fig. 28a) is shifted substantially towards the product side.

Table 47. The effect of exchange on the shapes of diabatic surfaces[a]

Item	Diabatic surface	LCFC-ZIDMOO		Rigorous LCFC	
		EE Arom. reaction	EE Antiarom. reaction	EE Antiarom. reaction	EE Arom. reaction
1	LU^D —— —— LU^A HO^D HO^A D A	R[b]	R[b]	R[c] (steep incline)	R[c] (small incline)
2	—— D$^+$ A$^-$	AR	AR	AR (higher and looser minimum)	AR (lower and tighter minimum)
3	—— D A*	R[b]	R[b]	R[c]	R[c]
		00 Nonarom. reaction		00 Nonarom. reaction	
4	LU^D —— —— LU^A HO^D HO^A D A	R[d]		AR	

269

Table 47 (continued)

			00 Nonarom. reaction	00 Nonarom. reaction
5	D$^+$	A$^-$	AR	AR
6	D	A*	R^d	AR

---

[a] R = Repulsive, AR = Attractive-Repulsive.

[b] Exaggerated minimum may exist at long interfragment distance.

[c] Shallow minimum may exist at long interfragment distance.

[d] Rigorous LCFC minimum may be reduced drastically and simultaneously shifted to long interfragment distance.

Epilogue

In 1965, Woodward and Hoffmann suggested that a correlation exists between MO properties and certain non-intuitive reactivity trends [3]. Subsequently, Woodward and Hoffmann [4] as well as Longuet-Higgins and Abrahamson [9] developed a sound theoretical basis for the treatment of many aspects of chemical reactivity. The conveyor of their theoretical ideas has been the correlation diagram, a close relative of the very familiar united atom correlation diagram. The key feature of their work has been the principle of conservation of orbital symmetry. Quite justifiedly, the Woodward-Hoffmann approach soon became a centerpiece of organic theory combining the elegance of simple theoretical notions and the brilliance required for correlating chemical facts and abstract ideas. Herein lies the only problem which I could identify with the socalled "Woodward-Hoffmann rules". These rules proved to be very powerful and initiated an unfortunate brainwashing cycle of experimental organic chemists who came to view organic reactions as Woodward-Hoffmann "allowed" or "forbidden".

An onslaught of theoretical publications followed, most of them purporting to do things "more correctly", "more adequately", "more generally", etc. In addition, criticism of the Woodward-Hoffmann approach was often promulgated based upon strawman constructs. In my opinion, these additional theoretical efforts have been helpful but cannot claim superiority, at least in any substantive way, over the original Woodward-Hoffmann approach.

In the early 1970's, I suggested that a correlation exists between the donor-acceptor matching of two molecules and various reactivity trends [6, 7]. As the donor-acceptor relationship of two reactants is progressively altered, different MO interactions become dominant and, as a result, a change of reaction choroselectivity may occur. The original approach was correlative to the extent that OEMO theory was used to assess orbital interactions within the principal resonance contributor of the reaction complex resonance hybrid. In addition, it was based on a number of assumptions. Nonetheless, many predicted trends correlated nicely with experimental facts and this provided hope for an eventual "rigorous" general theory of chemical reactivity. Still, a basic problem remained: the static model employed was known to suffer from key drawbacks. The reader can determine that the conclusions of this work differ not only from those of others but also from certain previous conclusions of ours based on the inferior static and correlation diagram models.

Scientific contributions are not entered in vacuo and, hence, they are always based on or inspired by previous knowledge. This work is no exception to this rule.

In particular, the static LCFC method has been utilized in attempts to interpret the electronic spectra of molecules and molecular complexes. Early attempts to apply this theoretical method to problems of chemical reactivity were made by R. D. Brown [145] who sought to interpret various aspects of electrophilic aromatic substitution. Turning our attention to the more fundamental problem of the construction of P.E. surfaces on the basis of the LCFC method, it is possible to identify the work of Mulliken and his collaborators on P.E. surfaces for molecular complex formation [22a] and the work of Nagakura [146] on P.E. surfaces for thermal electrophilic aromatic substitution reactions. Also, it should be pointed out that the LCFC method has been a very potent tool of spectroscopists and many successful applications to problems involving molecular electronic excitation can be found in the literature.

The advancement of our ideas to the chemical forum has not been unimpeded. Indeed, this work would never materialize had this author paid attention to suggestions of "expert" referees. In the space below, I present typical examples of such suggestions:

a) The rates of thermal Woodward-Hoffmann "forbidden" reactions do *not* depend on polarity but *only* on the size of the two electron interaction matrix elements of the ground and diexcited configurations.

b) As polarity increases, the stabilization of a Woodward-Hoffmann "allowed" sigmatropic shift *decreases.*

c) All Woodward-Hoffmann "allowed" reactions do *not* involve intermediates.

I hope that I have convinced the reader that, in each of the above cases, the opposite is true.

At this point, I would like to direct the attention of the reader to certain recent important contributions of other workers towards the understanding of chemical reactivity. Thus, in addition to the Woodward-Hoffmann contribution, the work of Herndon [10] is of pivotal importance insofar as it pointed out how simple OEMO theory can give a simple account of certain trends of photochemical reactions. Strangely, this work, which expresses ideas similar to the ones described in my initial publications, has not yet received the attention it deserves. The work of Houk on regioselectivity [27b, 28f], carried out independently of my own work on the same subject and encompassing a diverse range of topics, is also an important contribution. Despite specific disagreements, the theoretical discussions of certain aspects of photoreactions by Michl [142], Salem [147], Devaquet [148], and others [149], constitute interesting attempts to make sense out of the chaos of organic photochemistry. Finally, recent interesting contributions by Fueno et al. [150], Fukui [151], Goddard [152], and Pearson [153] deserve mention.

In summary, I have presented a discussion of organic reactivity based on two elementary diabatic surface shapes, the no bond type and the charge transfer type. In many cases, the conclusions have posed a challenge to currently accepted ideas, e.g., Chap. 6. In many other cases, new concepts have been developed, e.g., Chap. 12. Most importantly, certain problems were formulated but solution has not been possible, e.g., Chap. 20. The final interlinkage of theoretical ideas and experimental facts seems satisfactory. A good grasp of the main features of organic reactions may become a realistic possibility when the essential ideas of this work are further refined. Only time will tell whether these last projections have been overoptimistic. At present, however, one thing is clear: physical organic chemistry is alive and well and entering a new era when old taboos will crumble and a new overview of organic reactivity along the lines suggested in this work will emerge.

In conclusion, I wish to thank a number of people for the contribution they have made towards the realization of the present work. Special thanks are due to Professor P. von R. Schleyer who tolerated me as a graduate student at Princeton University, read part of this work and offered helpful suggestions. Professors K. Mislow, N. J. Turro, and L. Salem provided needed encouragement during the initial phases of this research. My predoctoral associates, especially Mr. S. Shaik, have assisted menially as well as intellectually. J. Holman, M. Palof and B. Jaeger converted an awfully hand-written text to the final printed version. Linda Daniel tolerated and encouraged an author who is not exactly fond of written communication. Finally, I wish to express my appreciation for the invitation extended by Dr. Boschke to contribute to this series and his efforts towards an excellent production of the work.

Appendix

Pi HO and LU Electron densities of organic molecules[a]

A structure diagram:

$$Y_1\!-\!\underset{Y_2}{\overset{Y_3}{C_\alpha}}\!=\!C_\beta\!-\!Y_4$$

Y$_1$	Y$_2$	Y$_3$	Y$_4$	Computation	HO Electron density C$_\alpha$	C$_\beta$	LU Electron density C$_\alpha$	C$_\beta$
F	H	H	H	CNDO/2	0.3014	0.4720	0.5132	0.4625
				ab initio	0.2639	0.4123	0.6829	0.6281
Cl	H	H	H	CNDO/2	0.1172	0.2546	0.5037	0.4784
				ab initio	0.2255	0.3011	0.6603	0.6703
OH	H	H	H	CNDO/2	0.2053	0.4402	0.5264	0.4390
				ab initio	0.1928	0.4091	0.7066	0.6005
SH	H	H	H	CNDO/2	0.0526	0.1522	0.5074	0.4783
				ab initio	0.0787	0.2338	0.6917	0.6268
CH$_3$	H	H	H	CNDO/2	0.3047	0.4500	0.4163	0.4448
				ab initio	0.3569	0.4088	0.6483	0.6376
CN	H	H	H	CNDO/2	0.2364	0.3546	0.2908	0.4319
				ab initio	0.2654	0.3021	0.3522	0.5553
NO$_2$	H	H	H	CNDO/2	0.5050	0.4560	0.4203	0.5243
CH$_2$=CH	H	H	H	CNDO/2	0.1706	0.3295	0.1810	0.3190
				ab initio	0.1604	0.2796	0.2143	0.3955
Ph	H	H	H	CNDO/2	0.0996	0.2404	0.1129	0.2352
F	F	H	H	CNDO/2	0.2375	0.5071	0.5269	0.4252
F	F	F	H	CNDO/2	0.2671	0.3689	0.4941	0.4413
F	F	F	Cl	CNDO/2	0.2545	0.2345	0.4732	0.4490
F	F	Cl	Cl	CNDO/2	0.2360	0.2097	0.4873	0.4390
F	F	F	CN	CNCO/2	0.2574	0.2852	0.4351	0.2623
F	F	CN	CN	CNDO/2	0.2256	0.2700	0.4670	0.2078
F	Cl	Cl	Cl	CNDO/2	0.2104	0.2030	0.4773	0.4526
Cl	Cl	Cl	H	CNDO/2	0.1640	0.2220	0.4880	0.4610

Appendix (continued)

B. $\overset{\alpha}{Y}\overset{\beta}{CH}=CH-\overset{\gamma}{CH}=\overset{\delta}{CH_2}$

Y	Computation	HO Electron density C_α	C_β	C_γ	C_δ	LU Electron density C_α	C_β	C_γ	C_δ
F	CNDO/2	0.2626	0.2931	0.1234	0.3055	0.3416	0.2962	0.1641	0.2995
	ab initio	0.2302	0.2035	0.1288	0.2554	0.4109	0.2020	0.2090	0.3840
Cl	CNDO/2	0.1979	0.2114	0.0801	0.2114	0.3342	0.3010	0.1651	0.3010
	ab initio	0.2486	0.1592	0.1396	0.2410	0.4104	0.2509	0.1781	0.3781
OH	CNDO/2	0.2191	0.2414	0.0932	0.2842	0.3579	0.1457	0.1769	0.2942
	ab initio	0.1951	0.2362	0.1061	0.2437	0.4202	0.1768	0.2231	0.3820
SH	CNDO/2	0.0583	0.0828	0.0180	0.0827[b]	0.3332	0.3005	0.1659	0.3005
		(0.235)	(0.280)	(0.190)	(0.280)				
CH₃	CNDO/2	0.3164	0.1966	0.1187	0.3134	0.3225	0.2900	0.1513	0.2905
	ab initio	0.2701	0.1857	0.1396	0.2692	0.3982	0.2042	0.2138	0.3864
CN	CNDO/2	0.3250	0.2465	0.1079	0.2465	0.2606	0.2506	0.1032	0.2506
	ab initio	0.2584	0.1430	0.1427	0.2359	0.3023	0.3079	0.1020	0.2979
NO₂	CNDO/2	~0.0	~0.0	~0.0	~0.0[b]	0.1706	0.1651	0.0424	0.1779
		(0.309)	(0.275)	(0.154)	(0.275)				
Ph	CNDO/2	0.1624	0.1726	0.0526	0.1726	0.3191	0.1809	0.0586	0.1651
H	CNDO/2	0.3244	0.1705	0.1705	0.3294	0.2146	0.2059	0.1809	0.3191
CH₂=CH	CNDO/2	0.2093	0.2156	0.0755	0.2166	0.2146	0.2059	0.0824	0.2071
	ab initio	0.2084	0.2083	0.0709	0.1810	0.2546	0.2546	0.0861	0.2476

a In CNDO/2 and Extended Hückel calculations of molecules containing only first row heteroatoms, standard bond angles and bond lengths have been used according to the prescription of ref. [18a]. Average experimental values were used for second row heteroatoms. The ab initio calculations were carried out with the Gaussian 70 series of programs (Hehre, W. J., Lathan, W. A., Ditchfield, R., Newton, M. D., Pople, J. A.: Quantum Chemistry Program Exchange, Indiana University, Bloomington, Program No. 236). In these computations, the C–Y bond lengths were optimized while the rest of the geometrical parameters were taken at their standard or average values. All ab initio computations made use of the STO-3G basis set. Finally, the electron densities at C_2 and C_3 of monosubstituted benzenes, where Y = OH, SH represent average values over two equivalent conformers. In all molecules containing second row heteroatoms, d orbitals were excluded.

b This MO has principal lone pair character. The electron densities shown in parenthesis belong to the pi MO next in decreasing energy order.

Appendix (continued)

Structural diagram: benzene ring with substituent X at position 1, Y at position 4, ring positions numbered 2, 3, 4, 5, 6.

X	Y	Computation	HO Electron density						LU Electron density					
			C_1	C_2	C_3	C_4	C_5	C_6	C_1	C_2	C_3	C_4	C_5	C_6
F	H	CNDO/2	0.2525	0.1060	0.0569	0.2922			0.0	0.2418	0.2580	0.0		
		ab initio	0.2243	0.0918	0.0531	0.2615			0.0	0.3173	0.3305	0.0		
Cl	H	CNDO/2	0.1334	0.0659	0.0256	0.1485			0.2576	0.0748	0.0578	0.2755		
		ab initio	0.2204	0.0711	0.0555	0.2383			0.4421	0.1205	0.0961	0.4186		
OH	H	CNDO/2	0.1775	0.1158	0.0364	0.2430			0.0	0.2410	0.2584	0.0		
		ab initio	0.1884	0.1043	0.0426	0.2460			0.0	0.3143	0.3320	0.0		
SH	H	CNDO/2	0.0520	0.0280	0.0095	0.0618			0.0548	0.0308	0.0103	0.0653		
		ab initio	0.1153	0.0823	0.0211	0.1555			0.0	0.3172	0.3297	0.0		
CN	H	CNDO/2	0.2383	0.0891	0.0485	0.2425[c]			0.2574	0.1038	0.0487	0.2621		
			(0.0	0.2531	0.2469	0.0)								
		ab initio	0.0	0.0636	0.0593	0.2376			0.3116	0.1411	0.0564	0.3324		
			(0.0	0.2148	0.2096	0.0)								
NO_2	H	CNDO/2	0.3162	0.0728	0.0781	0.2981[c]	0.1860	0.0311	0.1537	0.0946	0.0230	0.1775		
			(0.0	0.251	0.248	0.0)	0.0320	0.2460						
		ab initio	0.0	0.2137	0.2088	0.2716	~0.0	~0.0[b]	0.1717	0.1069	0.0261	0.2008		
			(0.2890	0.0544	0.0778	0.0)	(0.1070)	(0.0343)						
NO_2	2-OH	CNDO/2	0.0959	0.1284	0.0867	0.0237			0.0846	0.0761	0.0311	0.1656	0.0030	0.1230
NO_2	3-OH	CNDO/2	0.0464	0.1073	0.1874	0.1201			0.1284	0.1118	0.0128	0.1494	0.0240	0.0830
NO_2	4-OH	CNDO/2	~0.0	~0.0	~0.0	~0.0			0.1040	0.1033	0.0085	0.1590	0.0090	0.1010
			(0.2487)	(0.0323)	(0.1100)	(0.1700)								

[c] In nitrobenzene and benzonitrile, the benzenoid HO's are practically degenerate.

Appendix (continued)

D.

Y	X	Computation	HO Electron density				LU Electron density			
			C_1	C_2	C_3	C_4	C_1	C_2	C_3	C_4
H	H	Extended Hückel	0.212	0.070	–	–	0.212	0.080	–	–
OH	H		0.165	0.110	0.060	0.210	0.205	0.050	0.100	0.182
NO₂	H		0.058	0.002	0.037	0.036	0.017	0.180	0.007	0.202
NO₂	OH		0.070	0.001	0.053	0.030	0.035	0.174	0.030	0.190

E.

Y	X	Computation	HO Electron density		LU Electron density	
			C_9	C_{10}	C_9	C_{10}
H	H	Extended Hückel	0.181	0.181	0.221	0.221
CN	H		0.160	0.530	0.255	0.157
OH	H		0.077	0.213	0.185	0.182

F.

Computation	HO Electron density			LU Electron density		
	C_1	C_4	C_6	C_1	C_4	C_6
Extended Hückel	0.295	0.113	0.0	0.004	0.221	0.261

References

1. Wheland, G. W.: Resonance in Organic Chemistry. New York: John Wiley and Sons, Inc. 1955.
2. Heitler, W., London, F.: Z. Physik, *44,* 455 (1927); Slater, J. C.: Phys. Rev. *38,* 1109 (1931).
3. a) Woodward, R. B., Hoffmann, R.: J. Am. Chem. Soc. *87,* 395 (1965). b) Hoffmann, R., Woodward, R. B.: J. Am. Chem. Soc. *87,* 2046 (1965). c) Woodward, R. B., Hoffmann, R.: J. Am. Chem. Soc. *87,* 2511 (1965).
4. a) Woodward, R. B., Hoffmann, R.: The Conservation of Orbital Symmetry. Weinheim: Verlag Chemie 1970. b) Longuet-Higgins, H. C., Abrahamson, E.: J. Am. Chem. Soc. *87,* 2045 (1965). c) Oosterhoff, L. J.: private communication in Havinga, E., Schlatmann, J. L. M. A.: Tetrahedron *16,* 146 (1961). d) Fukui, K.: Acc. Chem. Res. *4,* 57 (1971). e) Zimmerman, H. E.: Acc. Chem. Res. *4,* 272 (1971). f) Dewar, M. J. S.: Angew. Chem., Int. Ed. Engl. *10,* 761 (1971). g) Evans, M. G.: Trans. Faraday Soc. *35,* 824 (1939).
5. Proskow, S., Simmons, H. E., Cairns, T. L.: J. Am. Chem. Soc. *88,* 5254 (1966).
6. a) Epiotis, N. D.: J. Am. Chem. Soc. *94,* 1924 (1972). b) Epiotis, N. D.: J. Am. Chem. Soc. *94,* 1935 (1972). c) Epiotis, N. D.: J. Am. Chem. Soc. *94,* 1941 (1972). d) Epiotis, N. D.: J. Am. Chem. Soc. *94,* 1946 (1972). e) Epiotis, N. D.: Angew. Chemie, Int. Ed. Engl. *13,* 751 (1974).
7. a) Epiotis, N. D.: J. Am. Chem. Soc. *95,* 1191 (1973). b) Epiotis, N. D.: J. Am. Chem. Soc. *95,* 1200 (1973). c) Epiotis, N. D.: J. Am. Chem. Soc. *95,* 1206 (1973). d) Epiotis, N. D.: J. Am. Chem. Soc. *95,* 1214 (1973).
8. Dewar, M. J. S.: The Molecular Orbital Theory of Organic Chemistry. New York: McGraw-Hill 1969.
9. Chemical Reactivity and Reaction Paths (ed., G. Klopman), Chap. 3, 4, 5 and 7. New York: Wiley-Interscience 1974.
10. a) Herndon, W. C.: Chem. Revs. *72,* 157 (1972). b) see [9], Chap. 7.
11. Fukui, K.: Fort. Chem. Forsch. *15,* 1 (1970).
12. Klopman, G.: J. Am. Chem. Soc. *90,* 223 (1968).
13. Salem, L.: J. Am. Chem. Soc. *90,* 543, 553 (1968).
14. Hudson, R. F.: Angew. Chem., Int. Ed. Engl. *12,* 36 (1973).
15. a) Houk, K. N.: Acc. Chem. Res. *8,* 361 (1975). b) Houk, K. N.: in press.
16. a) Kauzmann, W.: Quantum Chemistry. New York: Academic Press, Inc. 1957. b) Turner, A. G.: Methods in Molecular Orbital Theory. Englewood Cliffs, New Jersey: Prentice-Hall 1974. c) Borden, W. T.: Modern Molecular Orbital Theory for Organic Chemists. Englewood Cliffs: Prentice-Hall 1975.
17. a) Pople, J. A.: Acc. Chem. Res. *3,* 217 (1970). b) Hehre, W. J., Ditchfield, R., Radom, L., Pople, J. A.: J. Am. Chem. Soc. *92,* 4796 (1970). c) Schaeffer III, H. F.: The Electronic Structure of Atoms and Molecules. Menlo Park: Addison-Wesley 1972.
18. a) Pople, J. A., Beveridge, D. L.: Approximate Molecular Orbital Theory. New York: McGraw-Hill 1970. b) Klopman, G., O'Leary, B.: Fort. Chem. Forsch. *15,* 445 (1970). c) Murrell, J. N., Harget, A. J.: Semiempirical SCF-MO Theory of Molecules. New York: Wiley-Interscience 1972.

19. a) Hoffmann, R.: J. Chem. Phys. *39,* 1397 (1963). b) Hoffmann, R.: J. Chem. Phys. *40,* 2474 (1964).

20. a) Hückel, E.: Grundzüge der Theorie Ungesättigter und Aromatischer Verbindungen. Berlin: Verlag Chemie 1938. b) Streitwieser Jr., A.: Molecular Orbital Theory for Organic Chemists. New York: John Wiley and Sons 1961.

21. Epiotis, N. D., Cherry, W. R., Yates, R. L., Shaik, S., Bernardi, F.: Structural Theory of Organic Chemistry, in: Topics in Current Chemistry, Vol. 70, 1977.

22. The static LCFC method has been applied to many spectroscopic problems. a) Mulliken, R. S., Person, W. B.: Molecular Complexes. New York: Wiley-Interscience 1969. b) Murrell, J. N.: The Theory of Electronic Spectra of Organic Molecules. New York: Wiley 1963.

23. For related approximations employed in other computational schemes with different energy constants, see, *inter alia:* Hoffmann, R.: J. Chem. Phys. *40,* 2745 (1964); see [24b].

24. The reader can familiarize himself with the rules for the evaluation of matrix elements by reference to quantum chemistry texts. A particularly clear discussion is given, *inter alia,* in a) Sandorfy, C.: Electronic Spectra and Quantum Chemistry. Englewood Cliffs: Prentice Hall 1964, and b) McGlynn, S. P., Vanquickenborne, L. G., Kinoshita, M., Carroll, D. G.: Introduction to Applied Quantum Chemistry. New York: Holt, Rinehart, and Winston 1972.

25. a) Landau, L.: Phys. Z. Sowjet. *2,* 46 (1932). b) Zener, C.: Proc. Roy. Soc. *A137,* 6961 (1932). c) Robinson, G. W., Frosch, R. P.: J. Chem. Phys. *37,* 1962 (1962). d) Robinson, G. W., Frosch, R. P.: J. Chem. Phys. *38,* 1187 (1963). e) Jortner, J., Rice, S. A., Hochstrasser, R. M.: Advan. Photochem. *7,* 149 (1969). f) Jortner, J.: Pure Appl. Chem. *27,* 389 (1971).

26. Epiotis, N. D.: J. Am. Chem. Soc. *95,* 5624 (1973).

27. Houk, K. N.: J. Am. Chem. Soc. *94,* 8953 (1972).

28. For other important contributions to the problem of regioselectivity see: a) Feuer, J., Herndon, W. C., Hall, L. H.: Tetrahedron *24,* 2575 (1968). b) Eisenstein, O., Lefour, J. M., Anh, N. T.: Chem. Commun., *1971,* 969. c) Bastide, J., El Ghandour, N., Henri-Rousseau, O.: Tetrahedron Let. *1972,* 4225. d) Bastide, J., El Ghandour, N., and Henri-Rousseau, O.: Bull. Soc. Chim. France *1973,* 2290. e) Bastide, J., Henri-Rousseau, O.: Bull. Soc. Chim. France *1973,* 2290. e) Bastide, J., Henri-Rousseau, O.: Bull. Soc. Chim. France *1974,* 1037. f) Houk, K. N., Sims, J., Watts, C. R., Luskus, L. J.: J. Am. Chem. Soc. *95,* 7301 (1973).

29. For related discussions, see: a) [10]. b) [11]. c) [13]. d) Epiotis, N. D., Sarkanen, S., Bjorkquist, D., Bjorkquist, L., Yates, R.: J. Am. Chem. Soc. *96,* 4075 (1974).

30. a) Hudson, B. S., Kohler, B. E.: Chem. Phys. Let. *14,* 299 (1972). b) Dunning Jr., T. H., Hosteny, R. P., Shavitt, I.: J. Am. Chem. Soc. *95,* 5067 (1973). c) Campion, W. J., Karplus, M.: Mol. Phys. *25,* 921 (1973).

31. Bartlett, P. D.: Quart. Rev. *24,* 473 (1970).

32. a) Montgomery, L. K., Schueller, K., Bartlett, P. D.: J. Am. Chem. Soc. *86,* 622 (1964). b) Bartlett, P. D., Wallbillich, G. E. H.: ibid. *91,* 409 (1969).

33. Nishida, S., Moritani, I., Teraji, T.: J. Org. Chem. *38,* 1878 (1973).

34. Sarel, S., Felzenstein, A., Yovell, J.: Chem. Commun. *1973,* 859.

35. Wiberg, N., Buchler, J. W.: Chem. Ber. *96,* 3223 (1963); ibid. *97,* 618 (1964).

36. Huisgen, R., Steiner, G.: J. Am. Chem. Soc. *95,* 5054, 5055, 5056 (1973).

37. Huisgen, R., Steiner, G.: Tetrahedron Let. *1973,* 3763.

38. a) Huisgen, R., Schug, R., Steiner, G.: Angew. Chem. Int. Ed. Engl. *13,* 80, 81 (1974). b) Karle, I., Flippen, J., Huisgen, R., Schug, R.: J. Am. Chem. Soc. *97,* 5285 (1975).

39. a) Steiner, G., Huisgen, R.: Tetrahedron Let. *1973,* 3769. b) Fleischmann, F. K., Kelm, H.: Tetrahedron Let. *1973,* 3773.

40. Gompper, R.: Angew. Chem., Int. Ed. Engl. *8,* 312 (1969).

41. a) le Noble, W. J., Mukhtar, R.: J. Am. Chem. Soc. *97,* 5938 (1975). b) Paquette, L. A., Broadhurst, M. J., Read, L. K., Clardy, J.: J. Am. Chem. Soc. *95,* 4639 (1973).

42. Sustmann, R.: Pure Appl. Chem. *40,* 569 (1974).

43. Epiotis, N. D., Yates, R. L., Bernardi, F., Carlberg, D.: J. Am. Chem. Soc. *98,* 453 (1976).

44. Favini, G., Simonetta, M.: Theor. Chim. Acta *1,* 294 (1963).

45. Watanabe, K., Nakayama, T., Mottl, J.: J. Quant. Spect. Radiat. Transf. *2,* 369 (1962).

279

46. Bryce-Smith, D., Gilbert, A., Orger, B., Tyrvell, H.: Chem. Commun. *1974*, 334.
47. Lake, R. F., Thompson, H.: Proc. Roy. Soc. *A315*, 323 (1970).
48. Titov, Y. A.: Russ. Chem. Rev. *31*, 267 (1962).
49. Huisgen, R., Grashey, R., Sauer, J., in: The Chemistry of Alkenes (ed. S. Patai). New York: Interscience 1964.

50. See [48].
51. Mark, V.: J. Org. Chem. *39*, 3179 (1974).
52. Inukai, T., Kojima, T.: J. Org. Chem. *36*, 924 (1971).
53. a) Yates, P., Eaton, P.: J. Am. Chem. Soc. *82*, 4436 (1960). b) Fray, G. I., Robinson, R.: J. Am. Chem. Soc. *83*, 249 (1961). c) See [15]. d) Houk, K. N., Strozier, R. W.: J. Am. Chem. Soc. *95*, 4094 (1973), and reference cited therein.
54. Fleming, I., Karger, M. H.: J. Chem. Soc. (C) *1967*, 226.
55. Kiselev, V. D., Miller, J. G.: J. Am. Chem. Soc. *97*, 4036 (1975).
56. Singlet exciplexes: a) Saltiel, J., D'Agostino, J. T., Chapman, O. L., Lura, R. D.: J. Am. Chem. Soc. *93*, 2804 (1971). b) Mizuno, K., Pac, C., Sakurai, H.: J. Am. Chem. Soc. *96*, 2993 (1974). c) Caldwell, R. A., Smith, L.: J. Am. Chem. Soc. *96*, 2994 (1974). d) Creed, D., Caldwell, R. A.: J. Am. Chem. Soc. *96*, 7369 (1974). e) Creed, D., Wine, P. H., Caldwell, R. A., Melton, L. A.: J. Am. Chem. Soc. *98*, 621 (1976). f) Farid, S., Hartman, S. E., Doty, J. C., Williams, J. L. R.: J. Am. Chem. Soc. *97*, 3697 (1975). g) Yang, N. C., Shold, D. M., McVey, J. K.: J. Am. Chem. Soc. *97*, 5004 (1975). h) Yang, N. C., Srinivasachar, K., Kim, B., Libman, J.: J. Am. Chem. Soc. *97*, 5006 (1975). i) McCullough, J. J., Miller, R. C., Fung, D., Wu, W. S.: J. Am. Chem. Soc. *97*, 5942 (1975). j) Ferguson, J., Mau, A. W.-H.: Mol. Phys. *27*, 377 (1974). k) Ferguson, J., Miller, S. E. H.: Chem. Phys. Let. *36*, 635 (1975).
Triplet exciplexes: a) Caldwell, R. A.: J. Am. Chem. Soc. *95*, 1690 (1973). b) Caldwell, R. A., Smith, L.: J. Am. Chem. Soc. *96*, 2994 (1974). c) Caldwell, R. A., Sovocol, G. W., Gajewski, R. P.: J. Am. Chem. Soc. *95*, 2549 (1973). d) Farid, S., Hartman, S. E., Doty, J. C., Williams, J. L. R.: J. Am. Chem. Soc. *97*, 3697 (1975).
57. Birks, J. B.: Photophysics of Aromatic Molecules. New York: Wiley-Interscience 1970.
58. a) Huisgen, R., Feiler, L., Binsch, G.: Angew. Chemie, Int. Ed. Engl. *3*, 753 (1964). b) Martin, J. C., Goodlett, V. W., Burpitt, R. D.: J. Org. Chem. *30*, 4309 (1965). c) Montaigne, R., Ghosez, L.: Angew. Chemie, Int. Ed. Engl. *7*, 221 (1968).
59. Blomquist, A. T., Kwiatek, J.: J. Am. Chem. Soc. *73*, 2098 (1951).

60. Hopff, H., Rapp, W.: Chem. Abstr. *36*, 1614 (1942).
61. Pfleger, R., Jager, A.: Chem. Ber. *90*, 2460 (1957).
62. Rey, M., Roberts, S., Dieffenbacher, A., Dreiding, A. S.: Helv. Chim. Acta *53*, 417 (1970).
63. Brady, W. T., Roe Jr., R.: J. Am. Chem. Soc. *92*, 4618 (1970).
64. Brook, P. R., Harrison, J. M., Duke, A. J.: Chem. Commun. *1970*, 589.
65. De Selms, R. C., Delay, F.: J. Org. Chem. *37*, 2908 (1972).
66. Holder, R. W.: Ph. D. Dissertation, Yale University 1972.
67. Huisgen, R., Feiler, L., Binsch, G.: Angew. Chem., Int. Ed. Engl. *3*, 753 (1964).
68. Otto, P., Feiler, L. A., Husigen, R.: Angew. Chem., Int. Ed. Engl. *7*, 737 (1968).
69. Alder, K., v. Brachel, H.: Ann. Chem. *651*, 141 (1962).

70. Gibson, T. W., Erman, W. F.: J. Org. Chem. *37*, 1148 (1972).
71. Liu, R. S. H., Hammond, G. S.: J. Am. Chem. Soc. *89*, 4936 (1967).
72. Huisgen, R., Stangl, H., Sturrn, H. J., Wagenhofer, H.: Angew. Chem. *74*, 31 (1962).
73. Vrbaski, T., Cvetanović, R. J.: Can. J. Chem. *38*, 1053 (1960).
74. Charlton, J. L., de Mayo, P.: Can. J. Chem. *46*, 1041 (1968).
75. Charlton, J. L., Liao, C. C., De Mayo, P.: J. Am. Chem. Soc. *93*, 2463 (1971).
76. Saito, I., Takami, M., Matsuura, T.: Chem. Let. *1972*, 1195.
77. Smirnov-Zamkov, L. V., Piskovitina, G. A.: Ukr. Khim. Zh. *28*, 531 (1962).
78. Dewar, M. J. S., Fahey, R. C.: J. Am. Chem. Soc. *85*, 3645 (1963).
79. Poutsma, M. L.: J. Am. Chem. Soc. *87*, 2161 (1965).

80. Summerbell, R. K., Lunk, H. E.: J. Am. Chem. Soc. *79*, 4802 (1957).

81. a) de la Mare, P. B. D., Bolton, R.: Electrophilic Additions to Unsaturated Systems. New York: Elsevier 1966. b) Fahey, R. C.: Top. Stereochem. *3,* 237 (1968). Freeman, F.: Chem. Rev. *75,* 439 (1975).
82. Macoll, A.: in: The Chemistry of Alkenes (ed. S. Patai). New York: Interscience 1964.
83. Gassman, P. G.: Accounts Chem. Res. *4,* 128 (1971).
84. Cairncross, A., Blanchard Jr., E. P.: J. Am. Chem. Soc. *88,* 496 (1966).
85. For a discussion of correlation diagrams in relation to kinetic energy release, see: Smyth, K. C., Shannon, T. W.: J. Chem. Phys. *51,* 4633 (1969).
86. Williams, D. H., Hvistendahl, G.: J. Am. Chem. Soc. *96,* 6753 (1974).
87. For leading references, see: a) Hoytink, G. J.: in: Chemiluminescence and Bioluminescence (eds. M. J. Cormier, D. M. Hercules, and J. Lee), p. 147. New York: Plenum Press 1973. b) Weller, A., Zachariasse, K.: ibid., p. 181. c) Bard, A. J., Kreszthelyi, C. P., Tachikawa, H., Tokel, N. E.: ibid., p. 193. d) Weller, A.: in: 5th Nobel Symposium (ed. S. Claesson) p. 413. New York: Interscience 1967. e) Ottolenghi, M.: Acc. Chem. Res. *6,* 153 (1973). f) Gundermann, K.-D.: Topics Curr. Chem. *46,* 61 (1974). g) Chuang, T. J., Eisenthal, K. B.: J. Chem. Phys. *62,* 2213 (1975). h) Mataga, N.: in: The Exciplex (eds. M. Gordon and W. R. Ware), p. 113. New York: Academic Press 1975.
88. For reviews of experimental data, see: a) de la Mare, P. B. D., Ridd, J.: Aromatic Substitution – Nitration and Halogenation. New York: Academic Press 1959. b) Norman, R. O. C., Taylor, R.: Electrophilic Substitution in Benzenoid Compounds. New York: Elsevier 1965. c) Miller, J.: Nucleophilic Aromatic Substitution. Amsterdam: Elsevier 1968.
89. For the original frontier orbital treatment of aromatic substitutions, see: Fukui, K., Yonezawa, T., Shingu, H.: J. Chem. Phys. *20,* 722, 1434 (1952).
90. Olah, G. A.: Acc. Chem. Res. *4,* 240 (1971).
91. de Bie, D. A., Havinga, E.: Tetrahedron *21,* 2359 (1965).
92. Lodder, G., Havinga, E.: Tetrahedron *28,* 5583 (1972).
93. See, inter alia, [8].
94. T. H. Lowry and K. Schueller-Richardson: Mechanism and Theory in Organic Chemistry. New York: Harper and Row, 1976.
95. J. March: Advanced Organic Chemistry. New York: McGraw-Hill, 1977.
96. Letsinger, R. L., Hautala, R. R.: Tetrahedron Let. *1969.* 4205.
97. Peterson, W. C., Letsinger, R. L.: Tetrahedron Let. *1971,* 2197.
98. Lok, C. M. Havinga, E.: Proc. K. Ned. Akad. Wet. B *77,* 15 (1974).
99. Lammers, J. G.: Ph. D. Thesis, University of Leiden 1974.
100. Lok, C. M.: Ph. D. Thesis, University of Leiden 1972.
101. Cornelisse, J., Havinga, E.: Chem. Revs. *75,* 353 (1975).
102. Vink, J. A. J., Verheijt, P. L., Cornelisse, J., Havinga, E.: Tetrahedron *28,* 5081 (1972).
103. Reid, D. H., Stafford, W. H., Ward, J. P.: J. Chem. Soc. *1958,* 1100.
104. Anderson, A. G., Jr., Gale, D. J., McDonald, R. N., Anderson R. G., Rhodes, R. C.: J. Org. Chem. *29,* 1373 (1964).
105. See footnote on p. 141.
106. D. C. Nonhebel and J. C. Walton: Free Radical Chemistry. London: Cambridge University Press, 1974.
107. Spagnolo, S., Testaferri, L., Tiecco, M.: J. Chem. Soc. B *1971,* 2006.
108. Tiecco, M.: unpublished results.
109. Birchall, J. M.: unpublished results.
110. For related examples, see: Pryor, W. A., Davies, W. H., Jr., Gleaton, J. H.: J. Org. Chem. *40,* 2099 (1975).
111. Citterio, A., Minisci, F., Porta, O., Sesana, G.: J. Am. Chem. Soc. *99,* 7960 (1977).
112. Epiotis, N. D.: J. Am. Chem. Soc. *95,* 3188 (1973).
113. See also: a) Salem, L., Leforestier, C., Segal, G., Wetmore, R.: J. Am. Chem. Soc. *97,* 479 (1975). b) Salem, L., Stohrer, W. D.: Chem. Commun. *1975,* 140.
114. a) Douglas, J. E., Rabinovitch, B. S., Looney, F. S.: J. Chem. Phys. *23,* 315 (1955). b) Rabinovitch, B. S., Michel, K. W.: J. Am. Chem. Soc. *81,* 5065 (1959). c) Jones, J. L., Taylor, R. L.: J. Am. Chem. Soc. *62,* 3480 (1940).

115. Larkin, F. S.: Can. J. Chem. *46*, 1005 (1968).
116. Carlsson, D. J., Ingold, K. U.: J. Am. Chem. Soc. *90*, 7047 (1968); Burkhart, R. D., Boynton, R. F., Merrill, J. C.: J. Am. Chem. Soc. *93*, 5013 (1971).
117. Similar conclusions can be reached on the basis of OEMO theory. See: Berson, J. A., Salem, L.: J. Am. Chem. Soc. *94*, 8917 (1972).
118. Sneen, R. A.: Acc. Chem. Res. *6*, 46 (1973).
119. Berson, J. A., Holder, R.: cited in: Berson, J. A., Salem, L.: J. Am. Chem. Soc. *94*, 8917 (1972).
120. Klärner, F. G.: Tetrahedron Let. *1971*, 3611.
121. Cookson, R. C., Kemp, J. E.: Chem. Commun. *1971*, 385.
122. For a useful compilation of rate parameters of sigmatropic shifts, see: Willcott, M. R., Cargill, R. L., Sears, A. B.: Progress Phys. Org. Chem. *9*, 25 (1972).
123. Lewis, E. S., Hill, J. T., Newman, E. R.: J. Am. Chem. Soc. *90*, 662 (1968).
124. Evans, D. A., Golob, A. M.: J. Am. Chem. Soc. *97*, 4765 (1975).
125. Miller, B.: Acc. Chem. Res. *8*, 245 (1975).
126. Simpson, J. M., Richey Jr., H. G.: Tetrahedron Let. *1973*, 2545.
127. Spangler, C. W.: Chem. Revs. *76*, 187 (1976).
128. Cooper, W., Walters, W. D.: J. Am. Chem. Soc. *80*, 4220 (1958). Freedman, H. H., Doorakian, G. A., Sandel, V. R.: ibid. *87*, 3019 (1965).
129. Reinhoudt, D. N., Volger, H. C., Kouwenhoven, C. G., Wynberg, H., Helder, R.: Tetrahedron Let. *1972*, 5269.
130. Brannock, K. C., Burpitt, R. D., Goodlett, V. W., Thweatt, J. G.: J. Org. Chem. *28*, 1464 (1963).
131. Chapman, O. L., Pasto, D. J., Borden, G. W., Griswold, A. A.: J. Am. Chem. Soc. *84*, 1220 (1962).
132. a) Breslow, R., Napierski, J., Schmidt, A. H.: J. Am. Chem. Soc. *94*, 5906 (1972). b) Childs, R. F., Sakai, M., Parrington, B. D., Winstein, S.: J. Am. Chem. Soc. *96*, 6403 (1974) and references cited therein.
133. a) Schmidt, W.: Helv. Chim. Acta *54*, 862 (1971). b) Baldwin, J. E., Andrist, A. H., Pinschmidt Jr., R.: J. Am. Chem. Soc. *94*, 5845 (1972). c) Baldwin, J. E., Andrist, A. H., Pinschmidt Jr., R. K.: Acc. Chem. Res. *5*, 402 (1972).
134. a) Srinivasan, R.: Advan. Photochem. *4*, 113 (1968). b) Grimbert, D., Segal, G., Devaquet, A.: J. Am. Chem. Soc. *97*, 6629 (1975).
135. Turro, N. J., Ramamurthy, V.: Tetrahedron Let. *1976*, 2423.
136. Peyerimhoff, S. D., Buenker, R. J.: Advan. Quant. Chem. *9*, 69 (1975).
137. Sandorfy, C.: in: Progress in Theoretical Organic Chemistry, Vol. II (ed. I. G. Csizmadia). Amsterdam: Elsevier 1977.
138. Kornblum, N., Smiley, R. A., Blackwood, R. K., Iffland, D. C.: J. Am. Chem. Soc. *77*, 6269 (1955).
139. a) Kornblum, N., Seltzer, R., Haberfield, P.: ibid. *85*, 1148 (1963); Kornblum, N., Berrigan, P. J., le Noble, W. J.: ibid. *85*, 1141 (1963). b) Kornblum, N., Hardies, D. E.: ibid. *88*, 1701 (1966); Kornblum, N., Jones, W. J., Hardies, D. E.: ibid. *88*, 1704 (1966); Kornblum, N., Lurie, A. P.: ibid. *81*, 2705 (1959).
140. Dewar, M. J. S., Kirschner, S., Kollmar, K. W.: J. Am. Chem. Soc. *96*, 5240 (1974).
141. Hammond, G. S.: J. Am. Chem. Soc. *77*, 334 (1955).
142. a) Michl, J.: Topics Curr. Chem. *46*, 1 (1974), and reference therein. b) Michl, J.: Pure Appl. Chem. *41*, 507 (1975), and references therein.
143. a) Kita, S., Fukui, K.: Bull. Chem. Soc. Japan *42*, 66 (1969). b) Fukui, K.: Acc. Chem. Res. *4*, 57 (1971).
144. Gerhartz, W., Poshusta, R. D., Michl, J.: J. Am. Chem. Soc. *98*, 6427 (1976).
145. Brown, R. D.: J. Chem. Soc. *1959*, 2224, 2232.
146. Nagakura, S.: Tetrahedron *19*, Suppl. 2, 361 (1963).
147. Correlation diagrams for photoreactions: a) Salem, L.: J. Am. Chem. Soc. *96*, 3486 (1974). b) Salem, L., Leforestier, C., Segal, G., Wetmore, R.: J. Am. Chem. Soc. *97*, 479 (1975).

c) Dauben, W. G., Salem, L., Turro, N. J.: Acc. Chem. Res. *8,* 41 (1975). d) Salem, L.: Science *191,* 822 (1976). Two center SO coupling: e) Salem, L., Rowland, C.: Angew. Chem., Int. Ed. Engl. *11,* 92 (1972). f) Salem, L.: Pure Appl. Chem. *33,* 317 (1973).

148. Devaquet, A.: Pure Appl. Chem. *41,* 455 (1975).
149. a) Zimmerman, H. E.: J. Am. Chem. Soc. *88,* 1564, 1566 (1966). b) Dauben, W. G., Cargill, R. L., Coates, R. M., Saltiel, J.: J. Am. Chem. Soc. *88,* 2742 (1966). c) Dougherty, R. C.: J. Am. Chem. Soc. *93,* 7187 (1971).
150. a) Fueno, T., Nagase, S., Tatsumi, K., Yamaguchi, K.: Theor. Chim. Acta *26,* 43 (1972). b) Nagase, S., Fueno, T.: ibid. *35,* 217 (1974). c) Nagase, S., Fueno, T.: ibid. *41,* 59 (1976). d) Tatsumi, K., Yamaguchi, K., Fueno, T.: Tetrahedron *31,* 2899 (1975), and previous papers in this series. e) Yamaguchi, K., Fueno, T., Fukutome, H.: Chem. Phys. Let. *22,* 461 (1973), and subsequent papers.
151. Fukui, K.: Theory of Orientation and Stereoselection. Heidelberg: Springer-Verlag 1975.
152. a) Goddard III, W. A.: J. Am. Chem. Soc. *92,* 7520 (1970). b) Goddard III, W. A.: ibid. *94,* 793 (1972). c) Goddard III, W. A., Dunning Jr., T. H., Hunt, W. J., Hay, P. J.: Acc. Chem. Res. *6,* 368 (1973).
153. Pearson, R. G.: Symmetry Rules for Chemical Reactions. New York: Wiley-Interscience 1976.

Author Index

Subject Index

Reactivity and Structure

Concepts in Organic Chemistry

Editors: K. Hafner, J.-M. Lehn, C. W. Rees,
P. v. Ragué Schleyer, B. M. Trost, R. Zahradník

Volume 1: J. Tsuji

Organic Synthesis by Means of Transition Metal Complexes

A Systematic Approach

4 tables. IX, 199 pages. 1975
ISBN 3-540-07227-6

Contens: Comparison of synthetic reactions by transition metal complexes with those by Grignard reagents. – Formation of σ-bond involving transition metals. – Reactivities of σ-bonds involving transition metals. – Insertion reactions. – Liberation of organic compounds from the σ-bonded complexes. – Cyclization reactions, and related reactions. – Concluding remarks.

Volume 2: K. Fukui

Theory of Orientation and Stereoselection

72 figures, 2 tables. VII, 134 pages. 1975
ISBN 3-540-07426-0

Contens: Molecular Orbitals. – Chemical Reactivity Theory. – Interaction of Two Reacting Species. – Principles Governing the Reaction Pathway. – General Orientation Rule. – Reactivity Indices. – Various Examples. – Singlet-Triplet Selectivity. – Pseudoexcitation. – Three-species Interaction. – Orbital Catalysis. – Thermolytic Generation of Excited States. – Reaction Coordinate Formalism. – Correlation Diagram Approach. – The Nature of Chemical Reactions.
Appendix 1: Principles Governing the Reaction Path – An MO-Theoretical Interpretation. – Appendix 2: Orbital Interaction between Two Molecules.

Volume 3: H. Kwart, K. King

d-Orbitals in the Chemistry of Silicon, Phosphorus and Sulfur

Approx. 220 pages. 1977
ISBN 3-540-07953-X

Contents: Theoretical Basis for d-Orbital Involvement. – Physical Properties Related to dp-π Bonding. – The Effects of dp-π Bonding on Chemical Properties and Reactivity. – Pentacovalency.

Volume 4: W. P. Weber, G. W. Gokel

Phase Transfer Catalysis in Organic Synthesis

XV, 280 pages. 1977
ISBN 3-540-08377-4

Contents: Introduction and Principles. – The Reaction of Dichlorocarbene with Olefins. – Reactions of Dichlorocarbene with Non-Olefinic Substrates. – Dibromocarbene and Other Carbenes. – Synthesis of Ethers. – Synthesis of Esters. – Reactions of Cyanide Ion. – Reactions of Superoxide Ions. – Reactions of Other Hucleophiles. – Alkylation Reactions. – Oxidation Reactions. – Reduction Techniques. – Preparation and Reactions of Sulfur Containing Substrates. – Ylids. – Altered Reactivity. Addendum: Recent Developments in Phase Transfer Catalysis.

Volume 6: M. L. Bender, H. Komiyama

Cyclodextrin Chemistry

13 figures, 37 tables. Approx. 110 pages. 1977
ISBN 3-540-08577-7

Contents: Properties. – Inclusion Complex Formation. – Catalyses by Cyclodextrins Leading to Practical Usages of Cyclodextrins. – Covalent Catalyses. – Noncovalent Catalyses. – Asymmetric Catalyses by Cyclodextrins. – Improvement by Covalent and Monocovalent Modification.

Springer-Verlag
Berlin
Heidelberg
New York

Topics in Current Chemistry

Fortschritte der chemischen Forschung

Managing Editor: F. L. Boschke

Springer-Verlag
Berlin
Heidelberg
New York